BestMasters

Mit „BestMasters" zeichnet Springer die besten Masterarbeiten aus, die an renommierten Hochschulen in Deutschland, Österreich und der Schweiz entstanden sind. Die mit Höchstnote ausgezeichneten Arbeiten wurden durch Gutachter zur Veröffentlichung empfohlen und behandeln aktuelle Themen aus unterschiedlichen Fachgebieten der Naturwissenschaften, Psychologie, Technik und Wirtschaftswissenschaften. Die Reihe wendet sich an Praktiker und Wissenschaftler gleichermaßen und soll insbesondere auch Nachwuchswissenschaftlern Orientierung geben.

Springer awards "BestMasters" to the best master's theses which have been completed at renowned Universities in Germany, Austria, and Switzerland. The studies received highest marks and were recommended for publication by supervisors. They address current issues from various fields of research in natural sciences, psychology, technology, and economics. The series addresses practitioners as well as scientists and, in particular, offers guidance for early stage researchers.

More information about this series at http://www.springer.com/series/13198

Miran Lemmerer

Chemoselective Nucleophilic α-Amination of Amides

With a Foreword by Univ.-Prof. Dr. Nuno Maulide

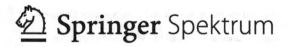 Springer Spektrum

Miran Lemmerer
Institute of Organic Chemistry
University of Vienna
Vienna, Austria

ISSN 2625-3577   ISSN 2625-3615   (electronic)
BestMasters
ISBN 978-3-658-30019-7   ISBN 978-3-658-30020-3   (eBook)
https://doi.org/10.1007/978-3-658-30020-3

This Springer Spektrum imprint is published by the registered company Springer Fachmedien
Wiesbaden GmbH part of Springer Nature.
The registered company address is: Abraham-Lincoln-Str. 46, 65189 Wiesbaden, Germany

Foreword

Contemporary science transcends the boundaries of traditional disciplines. Chemistry is an enabling and fundamental part of advances in physics, biology, material science and medicine to name but a few. One class of chemical compounds that is present in all living systems are so-called amides. Even though their importance is evident, they have often been overlooked as reactants in synthetic chemistry. Indeed, the remarkable stability of amides, a hallmark of their ubiquitous presence in biology, is reflected in their reluctance to engage with common reagents. Strikingly, a novel paradigm termed „electrophilic activation" allows one to make a "U-Turn" in reactivity – moving amides from a plateau of near-inertness to becoming one of the most reactive organic functional groups. The author of this book joined our venture in exploring this „amide activation" and developed, together with our team, a technique to synthesise a range of derivatives compounds including α-amino amides, a specific class of amides common in proteins. Clearly, the results presented here are likely to inspire not only other scholars in the field but especially practitioners of fields beyond chemistry.

Univ.-Prof. Dr. Nuno Maulide

Preface

First of all, I would like to thank Univ.-Prof. Dr. Nuno Maulide for giving me the chance to work on this exciting chemistry, for the excellent guidance throughout my thesis and for teaching me to always keep the big picture in mind.

Big thanks goes to Dr. Christopher Teskey for his continuous support in every tricky situation from the very first day in the lab until the last day in the office and for many inspiring discussions about chemistry.

Regarding the project itself, I would like to thank Dr. Pauline Adler and Carlos Goncalves as well as Dr. Daniel Kaiser for the fruitful conversations and their helpful suggestions.

Additionally, special thanks goes to all members of the Maulide group for the amazing time inside and outside of the university. Thanks to Martina Drescher, Mirjam Czuba, Elena Macoratti and Patricia Emberger for helping me, whenever I had a problem. Also, the big lab including Immo Klose, Giovanni Di Mauro, Carlos Goncalves, Bogdan Brutiu, Dr. Amandine Pons, Eric Lopes, Dr. Sébastien Lemouzy and many students need to be thanked for always radiating joy, even after a long day. Furthermore, thanks goes to my colleagues in the office: Samuel Senoner, Veronica Tona, Dr. Tobias Stopka, Yong Chen and Lorena Garcia for the great work environment and helping me through the less exciting parts of research.

Thanks to Dipl. Ing. Alexander Roller for the help with the crystal structure.

At last I would like to thank my friends and family for their exhaustless support!

The results of my master's thesis which is the basis of this book were partially published in an open access article. (Gonçalves, C. R.; Lemmerer, M.; Teskey, C. J.; Adler, P.; Kaiser, D.; Maryasin, B.; González, L.; Maulide, N. Unified Approach to the Chemoselective α-Functionalization of Amides with Heteroatom Nucleophiles *J. Am. Chem. Soc.* **2019**, *141*, 18437–18443.)

Miran Lemmerer

Table of Contents

Table of Abbreviations

Abbreviation	Meaning
Ar	aryl
Alk	alkyl
Boc	*tert*-butyloxycarbonyl
CAN	cerium ammonium nitrate
cat.	catalyst
18-crown-6	1,4,7,10,13,16-hexaoxacyclooctadecane
d	doublet
δ	chemical shift
DBU	1,8-Diazabicyclo[5.4.0]undec-7-ene
DCM	dichloromethane
DEAD	diethyl azodicarboxylate
DMF	*N,N*-dimethylformamide
d.r.	diastereomeric ratio
ee	enantiomeric excess
HMDS	bis(trimethylsilyl)amide
J	coupling constant
LN^+	2,6-lutidinium ion
LNO	2,6-lutidine *N*-oxide
m	multiplet
MeCN	acetonitrile
Ms	mesyl, methanesulfonyl
ν	wave number
*n*Bu	*n*-butyl
NHC	*N*-heterocyclic carbene
Ns	nosyl, 4-nitrobenzenesulfonyl

PMP	*para*-methoxyphenyl
Pr	propyl
q	quartet
r.t.	room temperature
s	singlet
t	triplet
TBAB	tetrabutylammonium bromide
TBAF	tetrabutylammonium fluoride
TBAHFP	tetrabutylammonium hexafluorophosphate
*t*Bu	*t*ert-butyl
TEBnAC	triethylbenzylammonium chloride
TEMPO	2,2,6,6-tetramethylpiperidin-1-yl)oxyl
Tf	triflyl, trifluoromethylsulfonyl
Ts	tosyl, toluensulfonyl

1 Introduction

1.1 Umpolung

By definition carbonyl moieties display electrophilic character at the carbonyl carbon and nucleophilic character at the α-position. Through deprotonation aldehydes, ketones, esters, amides and nitriles form enolates which react with electrophiles to yield α-functionalized products. Examples of such electrophiles include aldehydes, ketones or esters themselves. These aldol[1] or Claisen[2] coupling reactions showcase the electrophilic character of the carbonyl carbon as well as the nucleophilic character of the α-position.

To expand the number of possible transformations of carbonyl compounds, chemists have conceived creative strategies to reverse their polarity and therefore their reactivity. The term Umpolung ("polarity inversion") was popularized by the German chemist D. Seebach and the American Nobel laureate E. J. Corey and describes a reversal in partial charge.[3] Umpoled carbonyls are thus either nucleophilic at the carbonyl carbon or electrophilic at the α-position (Figure 1).

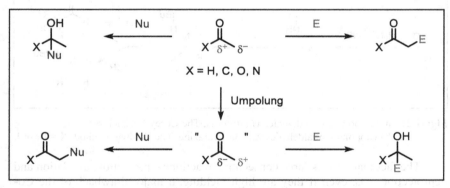

Figure 1: General depiction of the Umpolung of carbonyl groups.

In 1975, Seebach and Corey developed one of the first Umpolung protocols for aldehydes.[4] Through thioacetal formation, the original aldehydic hydrogen becomes more acidic than the α-proton and therefore enables selective deprotonation by a strong base such as *n*BuLi. The *in situ* formed (1,3-dithian-2-yl)lithium species (**I**) is nucleophilic at the previously electrophilic carbonyl carbon and reacts with electrophiles like aldehydes, ketones, alkyl halides and epoxides (Figure

© The Editor(s) (if applicable) and The Author(s), under exclusive license to Springer Fachmedien Wiesbaden GmbH, part of Springer Nature 2020
M. Lemmerer, *Chemoselective Nucleophilic α-Amination of Amides*, BestMaters, https://doi.org/10.1007/978-3-658-30020-3_1

2a). The thioacetal can be deprotected under oxidative conditions to reveal the desired ketone.[5] A similar method was employed by Baldwin et al. who used *tert*-butyl hydrazones instead of 1,3-dithianes.[6] These can be easily synthesised from aldehydes and *tert*-butyl hydrazine. Deprotonation with *n*BuLi affords an anionic acyl intermediate which can be reacted with an electrophile at carbon (Figure 2b). Deprotection can be achieved under acidic conditions.

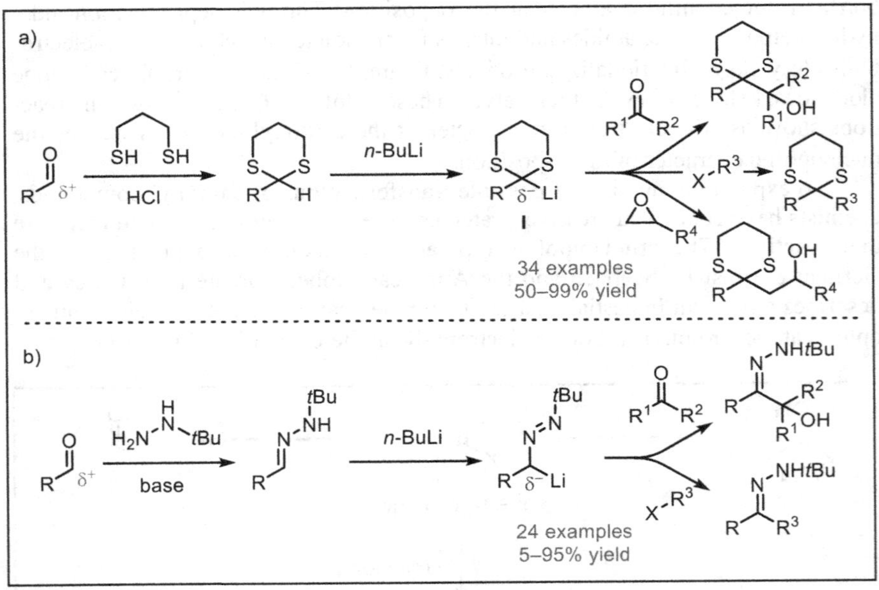

Figure 2:   a) Umpolung of aldehydes *via* dithiols: The Corey-Seebach reaction.
                 b) Umpolung of aldehydes *via* tBu-hydrazine: The Baldwin method. X = Br or I.

The fact that it takes three consecutive reactions – protection, alkylation and deprotection – is, even if they are high yielding, a major drawback of the Corey-Seebach or the Baldwin reactions. A more atom-economical approach is the Umpolung by cyanide. This concept was first discovered in the benzoin condensation.[7] Through the attack of cyanide on benzaldehyde, a nucleophilic intermediate is formed which immediately reacts with a second equivalent of benzaldehyde to form an α-hydroxy ketone and cyanide (Figure 3a). A more general approach was reported by Stetter in 1976.[8] He used an *N*-heterocyclic carbene (NHC) to form a nucleophilic species now known as the Breslow intermediate.[9] The latter, effectively an Umpoled aldehyde, can attack α,β-unsaturated ketones, esters or nitriles in a Michael addition (Figure 3b). Since the original report, many extensions of the Stetter reaction have been published including asymmetric variants.[10,11] The structure of the catalyst is derived from thiamine (Vitamin B$_1$). This compound acts as a coenzyme for pyruvate dehydrogenase, amongst other proteins, and is therefore essential for all organisms.[12]

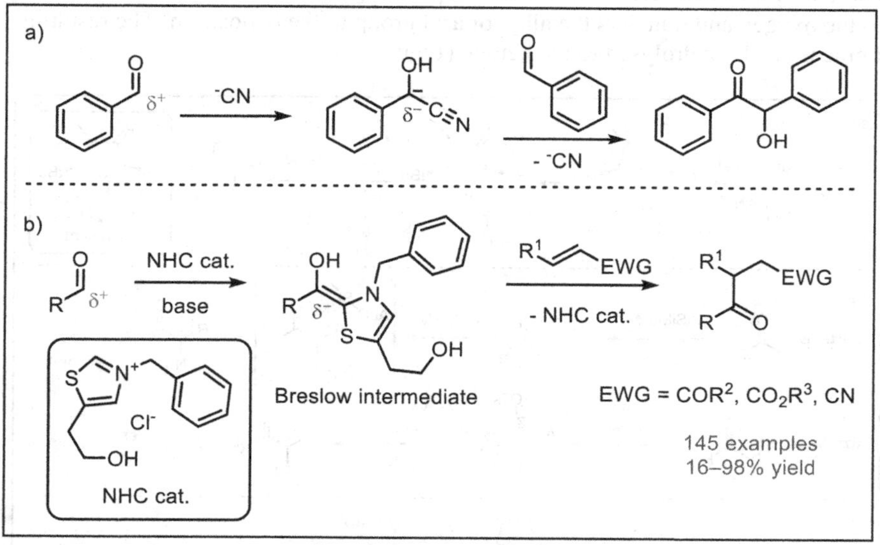

Figure 3: Catalytic Umpolung of aldehydes: a) The benzoin condensation, b) N-heterocyclic carbene catalysis.

While Umpolung at the carbonyl carbon is now a well-established concept, with many methods reported, Umpolung of the α-carbon is comparatively underexplored. In 2007 MacMillan *et al.* reported single electron oxidation of an *in situ* formed enamine to form an electrophilic α-carbon on aldehydes.[13] This method can be used to form α-allylated aldehydes enantioselectively (Figure 4a). In its racemic form, this concept was originally developed by Murakami *et al.* in 1992.[14]

Wirth *et al.* showed that treating ketones with a base and a silyl chloride containing an oxygen or nitrogen substituent yields a vinyl silyl ether (Figure 4b).[15] The reaction of this intermediate with an I^{III} species leads to *in situ* generation of an α-I^{III}-ketone which can be attacked by the nucleophilic silyl group substituent. The employment of a chiral iodine compound yields functionalized ketones with an *ee* up to 94%. One year after Wirth *et al.* had developed this method Szpilman *et al.* used Koser's reagent which is a similar hypervalent iodine compound in

combination with a nucleophilic alkyl zinc species to install an alkyl group in
α-position of a ketone (Figure 4b).[16]

In 2011, Miyata *et al.* developed the concept of using nucleophilic organo
aluminium compounds on isoxazolidine enamines.[17] These enamines are prepared *in situ* from a ketone. In the next step the organometallic species coordinates
to the oxygen and transfers the alkyl or aryl group to the α-position. The resulting
imine is easily hydrolysed to the ketone (Figure 4c).

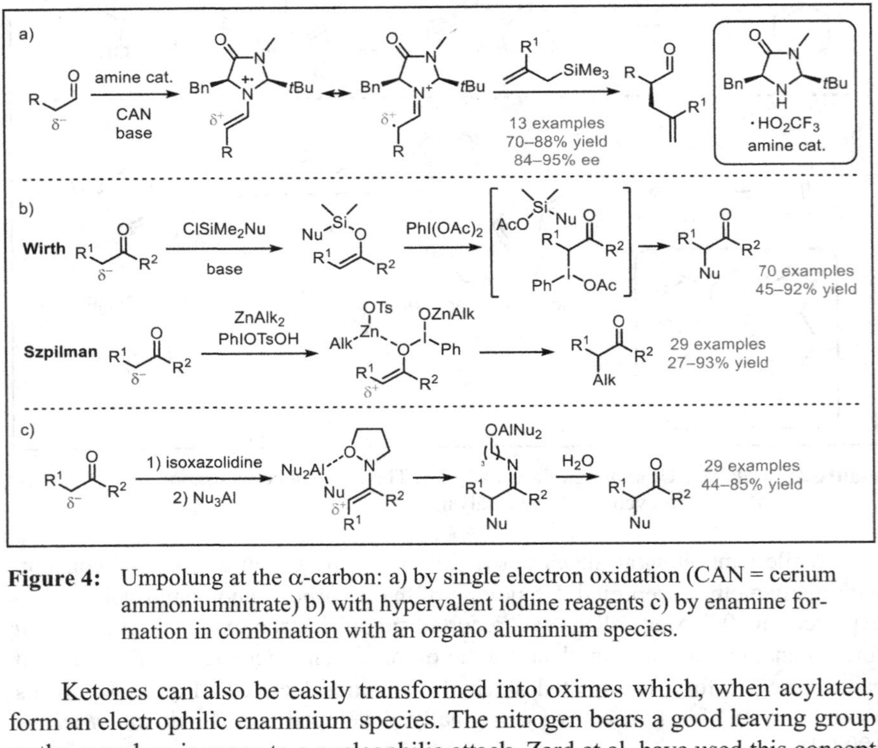

Figure 4: Umpolung at the α-carbon: a) by single electron oxidation (CAN = cerium
ammoniumnitrate) b) with hypervalent iodine reagents c) by enamine formation in combination with an organo aluminium species.

Ketones can also be easily transformed into oximes which, when acylated,
form an electrophilic enaminium species. The nitrogen bears a good leaving group
so the α-carbon is prone to a nucleophilic attack. Zard et al. have used this concept
to form tetracycles *via* Fridel Crafts type cyclisation in moderate yields (Figure
5).[18]

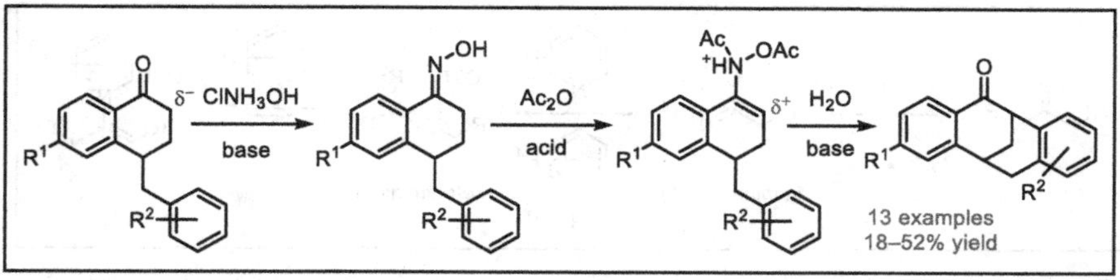

Figure 5: α-Umpolung of a ketone *via* sequential oxime formation and acetylation.

Maulide *et al.* have developed a new strategy for the chemoselective α-Umpolung of amides which will be described in the next chapter.

1.2 Amide Activation

Amides are ubiquitous functional groups in nature. In proteins, amino acids are connected *via* amide bonds. The unreactive nature of this bond is important for mechanical and chemical resistance in solution. The low electrophilicity of amides can be explained by delocalisation of the nitrogen lone pair into the π* orbital of the carbonyl.[19] This property however makes them more nucleophilic than most other carbonyl derivatives (Figure 6).

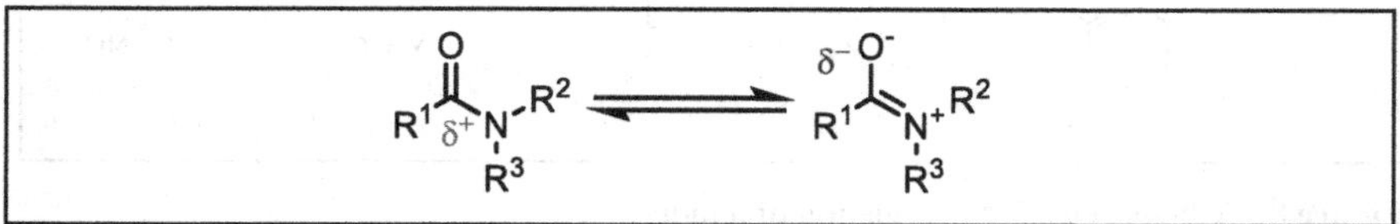

Figure 6: The mesomeric structures of amides.

Strong electrophiles can activate the amide carbonyl oxygen. One such example is trifluoromethanesulfonic anhydride (Tf$_2$O). Since triflate is a good leaving group, the iminium triflate formed in the reaction of an amide with Tf$_2$O can be deprotonated by weak bases to yield a highly electrophilic keteniminium ion.[20] Spectroscopic studies have shown that this intermediate is stabilized by a second equivalent of the pyridine derivative (Figure 7).[21]

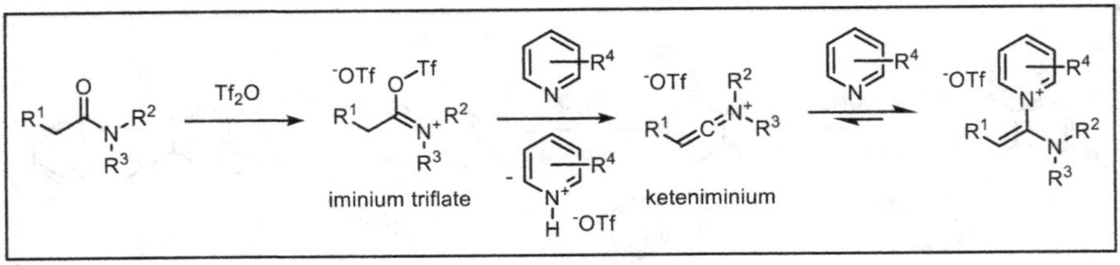

Figure 7: Formation of a keteniminium ion and stabilisation by a pyridine base. $R^4 = H$, CH_3, F, Cl, Br or I.

The afore mentioned nucleophilicity of amides enables their selective activation by triflic anhydride even in the presence of aldehydes, ketones, esters and nitriles. Many reactions have been developed using this method.[22] Notably, Maulide et al. discovered that aryl sulfoxides and aryl hydroxamic acids add to the keteniminium intermediate to trigger a [3,3]-sigmatropic rearrangement event, ultimately yielding α-arylated amide products (Figure 8).[23,24]

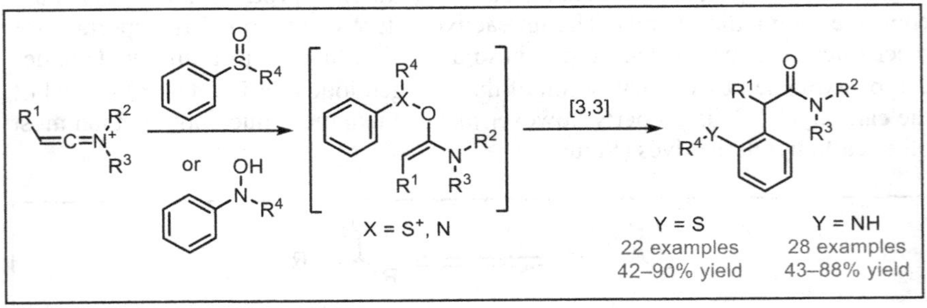

Figure 8: Chemoselective α-arylation of amides.

The chemoselective α-amination of amides can be achieved by the analogous addition of alkyl azides.[25] The unusual enamine intermediate **II** cyclises to a 2H-azirinium, the hydrolysis of which reveals the final products (Figure 9).

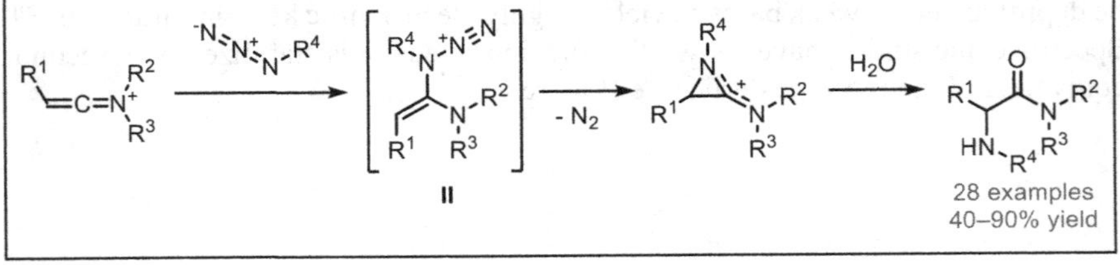

Figure 9: Chemoselective α-amination of amides.

In the same study, one special case was encountered. The use of a substrate bearing a tethered ester led to an unusual cyclisation to product **III**, *via* intermediate **IV** (Figure 10a). The addition of an alkyl azide, bearing a good leaving group (N_2), resulted in a polarity inversion at the α-carbon of the staring amide.

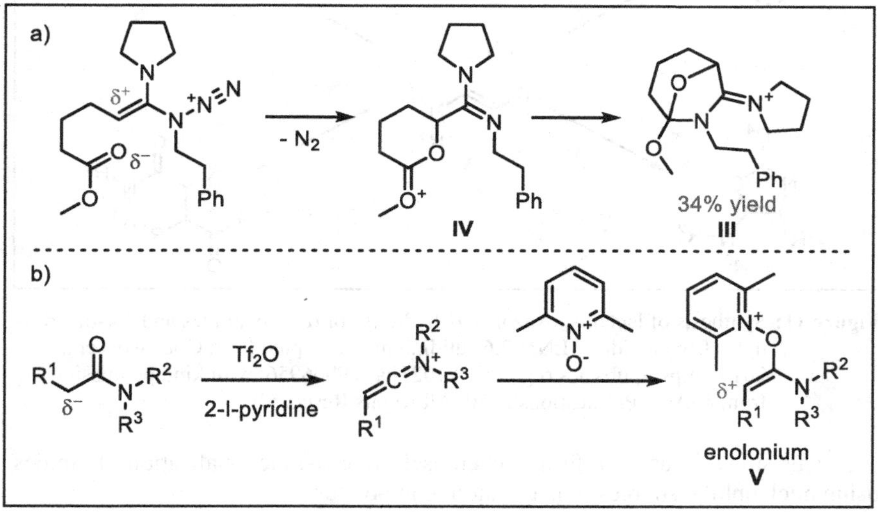

Figure 10: Umpolung of the α-carbon with a) an azide or b) 2,6-lutidine N-Oxide (LNO).

A 1979 report by Ghosez *et al.* provided an additional piece of evidence for this picture.[26] In that publication, Ghosez employed a pyridine *N*-oxide to intercept the keteniminium intermediate, leading to α,β-desaturation of the amide precursors. The involved intermediate **V** shows characteristics typical of an enolate structure whilst having noteworthy electrophilic character at the a-carbon. The term "enolonium" is thus particularly suitable for this species (Figure 10b).

The advent of this electrophilic enolonium intermediate has given rise to multiple new reactions. The first one reported was an intramolecular Fridel Crafts type cyclisation yielding 1,4-dihydroisoquinolin-3-one amongst other lactams.[27] Later on, oxazoles were synthesised *via* an attack of a nitrile,[28] α-oxidation was achieved through employment of a second equivalent 2,6-lutidine *N*-Oxide (LNO)[29] and 1,4-dicarbonyles were the products of the reaction of enolonium ions coupled with enolates (Figure 11).[30]

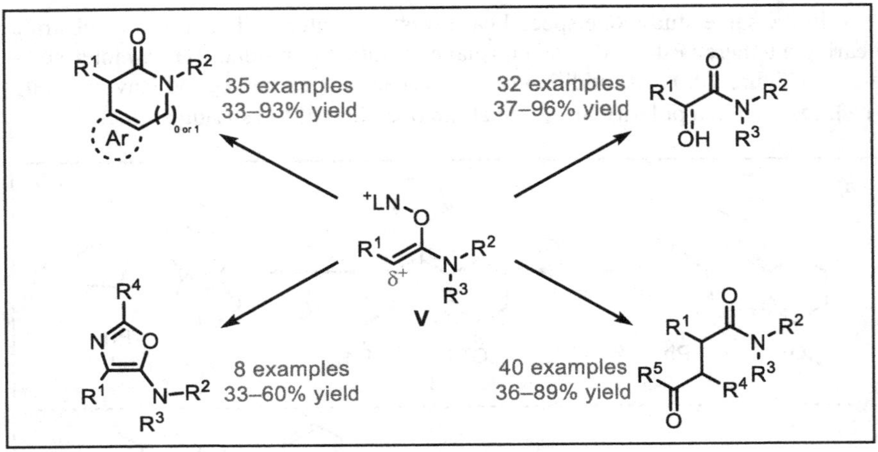

Figure 11: Synthesis of lactams, oxazoles, α-hydroxy- or α-keto amides and 1,4-dicarbo-
nyles from amides. $^+$LN = 2,6-lutidinium. (Addapted from Gonçalves et al.
2019, https://pubs.acs.org/doi/10.1021/jacs.9b06956; with kind permission
from © ACS Publications 2020. All Rights Reserved)

This study focuses on further chemoselective α-functionalisation of amides
using nucleophilic sources of heteroatom compounds.

1.3   Smiles Rearrangement

The Smiles rearrangement is a powerful tool in the construction of arene-bearing molecular structures.[31] Its reaction mechanism proceeds through a nucleophilic attack onto a (usually electron-poor) arene *ipso*- to a good leaving group. Particularly interesting is the α-arylation of amino acid derivatives. The deprotonation of nitrobenzenesulfonyl (nosyl)- protected amino acid esters or amides leads to the formation of a transient enolate which undergoes attack to the aromatic group to form a spirocyclic Meisenheimer intermediate.[32] With the irreversible loss of SO_2, the rearranged product is formed (Figure 12). Esters have been reported to undergo this rearrangement in good yields, and Wilson *et al.* synthesised 11 different α-nitro phenyl amino acid esters in up to 81% yield.[33] They show that aqueous *tetra*-butyl ammonium hydroxide as a base gives the best results. Later on, Lupi *et al.* have shown that the reaction proceeds in a stereospecific manner.[34] The first and only report of amides undergoing this rearrangement have been developed on compounds connected to a solid phase *via* a linker.[35] Only three examples of amides undergoing this transformation have been published.

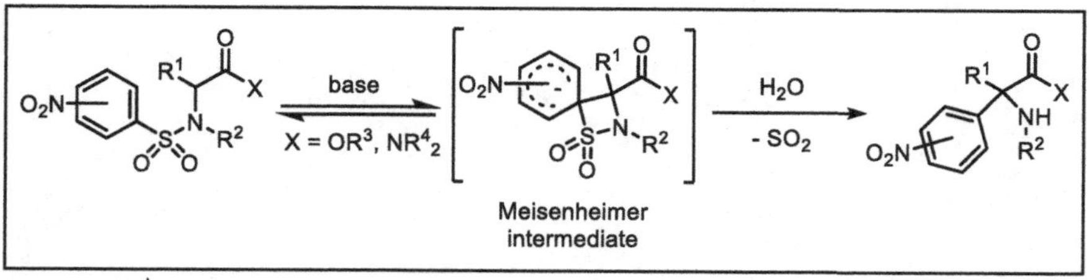

Meisenheimer
intermediate

Figure 12: α-Arylation of esters and amides *via* a Smiles rearrangement.

This specific Smiles chemistry will be further explored in the following chapters.

2 Results and Discussion

2.1 Nucleophilic α-Functionalisation of Amides

We envisioned a chemoselective nucleophilic functionalisation of amides through Umpolung. A challenge that could arise are the different electrophilic centres on the enolonium ion **V** itself. Attack on the α-carbon would lead to the desired product, while the addition of the nucleophile at carbon 4 of the lutidinium moiety would lead to a functionalised lutidine with concomitant release of the starting (Figure 13). In prior work we have shown that Tf₂O can activate pyridine *N*-oxides in that fashion.[36]

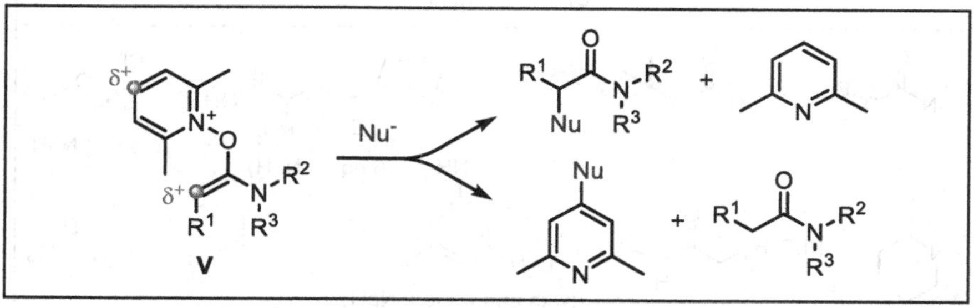

Figure 13: The different electrophilic sites on the enolonium. (Adapted from Gonçalves et al. 2019, https://pubs.acs.org/doi/10.1021/jacs.9b06956; with kind permission from © ACS Publications 2020. All Rights Reserved)

Despite this concern, we realised that the successful addition of nitrogen nucleophiles would result in an access to complex amino acid derivatives.

2.1.1 Amination

At the outset, we considered a range of possible nitrogen nucleophiles. The enolonium species was generated, as in prior studies, in dichloromethane (DCM) at 0 °C by the addition of LNO to the previously generated keteniminium ion.[30] Deprotonation of the nitrogen nucleophile was achieved with sodium hydride (NaH) in *N,N*-dimethyl formamide (DMF). This solution was then added to the enolonium intermediate and warmed to room temperature for 1 h. Various nitrogen nucleophiles were investigated including carbamates, indole, unprotected

© The Editor(s) (if applicable) and The Author(s), under exclusive license to Springer Fachmedien Wiesbaden GmbH, part of Springer Nature 2020
M. Lemmerer, *Chemoselective Nucleophilic α-Amination of Amides*, BestMaters, https://doi.org/10.1007/978-3-658-30020-3_2

tryptophan, dimethylhydroxylamine, amides, aniline derivatives and tosylhydra-zone. From all these nucleophiles only indole yielded the desired product **2.1a** while the others led to degradation or reformation of the amide (Figure 14). Another approach was to add metal amides directly. For this, lithium bis(trimethylsi-lyl)amide (LiHMDS), NaNH$_2$ or potassium phthalimidate were added to the enolonium solution. These also failed to aminate the amide and did not react with the enolonium ion.

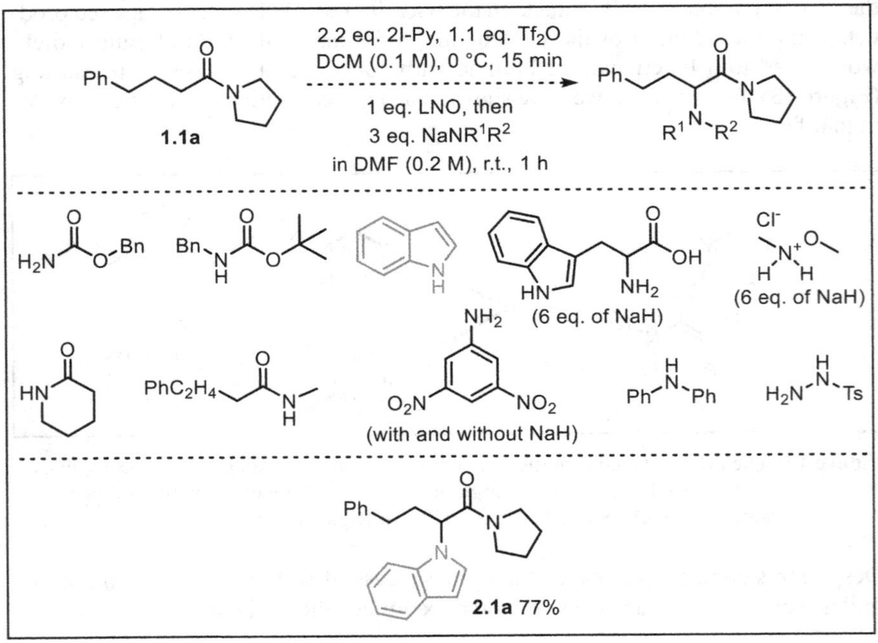

Figure 14: Nucleophilic amination attempts.

Deprotonated sulfonamides, however, readily underwent the desired trans-formation and yielded protected α-aminated amides. Methyl *para*-nitrophenyl-sulfonyl (nosyl or Ns) amide **1.2a** gave product **2.1b** in 82% isolated yield. After this encouraging result, some optimisation for the deprotonation and addition of the sulfonamide was conducted (Table 1). Direct addition of the sulfonamide with-out a base or sequentially with a base or a phase transfer catalyst did not give the product (entry 1–4). Switching the solvent to acetonitrile (MeCN) lowered the yield (entry 5–10). After this study, it was determined that the initial conditions were the most suitable and the preparation of the sodium salt in DMF was neces-sary.

Table 1: Control experiments for the amination with methyl Ns amide. TBAHFP = Tetrabutylammonium hexafluorophosphate. NMR yields were determined with 1,3,5-trimethoxybenzene as an internal standard. (Addapted from Gonçalves et al. 2019, https://pubs.acs.org/doi/10.1021/jacs.9b06956; with kind permission from © ACS Publications 2020. All Rights Reserved)

entry	T (°C)	Time (h)	solvent	Nucleophile addition conditions	NMR yield
1	r.t.	1	DCM	no base	0
2	r.t.	1	DCM	NaH + sulfonamide added directly	0
3	r.t.	1	DCM	Cs_2CO_3 + sulfonamide added directly	0
4	r.t.	1	DCM	no base + TBAHFP	0
5	r.t.	1	MeCN	NaH + sulfonamide added as a MeCN suspension	37
6	r.t.	1	MeCN	direct addition of Na-sulfonamidate	44
7	r.t.	20	MeCN	direct addition of Na-sulfonamidate	43
8	r.t.	2	MeCN	KHMDS + sulfonamide added directly	25
9	r.t.	2	MeCN	Na-sulfonamidate dissolved in 0.5 mL DMF	45
10	80	2	MeCN	direct addition of Na-sulfonamidate salt	31

To investigate the influence of the substituent on the nitrogen, a range of *para*-methylphenylsulfonyl (tosyl or Ts) amides were synthesised. These, as well as some commercially available nosyl- and tosyl-amides, were tested under the reaction conditions described above. Methyl-tosyl-amide gave product **2.1c** in 71% yield. Unsubstituted compound **2.1d** was synthesised in 69% yield which is comparable to the 71% of the sterically demanding isopropyl amide **2.1e**. A benzyl

(2.1f) as well as an unsubstituted phenyl **(2.1g)** and a *para*-methoxy phenyl **(2.1h)** were tolerated. We were pleased to see that tosyl- protected glycine methyl ester gave coupled product **2.1i** in a good 63% yield (Figure 15). The possibility of coupling *N*-tosylated amino acids opens up interesting applications.

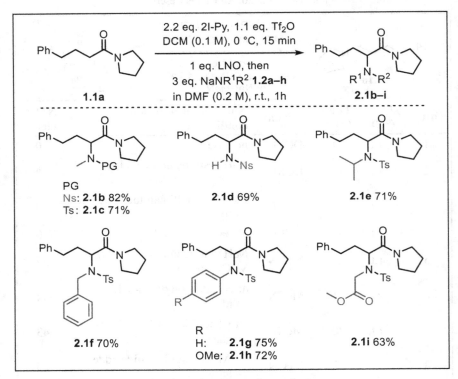

Figure 15: Addition of tosyl and nosyl amides to the enolonium.

Since tosyl- and nosyl- sulfonamides proved to be suitable nucleophiles, in this transformation, we further investigated other aromatic groups on the sulfona-mide. An *ortho*-nitrophenylsulfonyl amide gave the aminated product **2.1j** in a similar yield as its isomer **2.1b** while a naphthyl derivative **2.1k** was comparable to **2.1c**. The 2-pyridinyl sulfonamide **2.1l** was successfully obtained while its 4-ni-tro derivative **2.1m** gave no product. In the last case, the sulfonamide may have undergone decomposition and/or deprotection as none of it could be detected after the reaction (Figure 16).

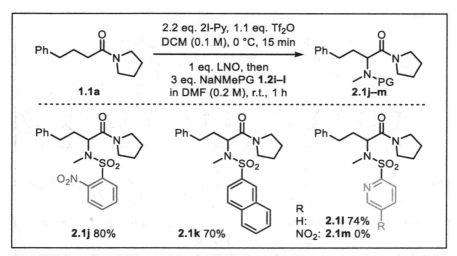

Figure 16: Different aromatic groups on the sulfonamide.

Our group and others have previously shown that the activation of amides with Tf₂O and a pyridine base is very chemoselective over other carbonyl groups.[22] In our case esters (**2.1n**), ketones (**2.1o**) and nitriles (**2.1p**) were tolerated. Pleasingly, a primary chloride was not substituted by the sulfonamide and instead smoothly yielded the desired product **2.1q**. A tethered alkene did not trigger a [2+2] cycloaddition[37] but rather gave amide **2.1r**. An amide derived from dimethylamine instead of pyrolidine could also be α-aminated (**2.1s**). Products **2.1u** and **2.1v** again showcased the similarity in the employment of tosyl and nosyl nucleophiles, while amides **2.1w** and **2.1x** could not be prepared *via* this method (Figure 17).

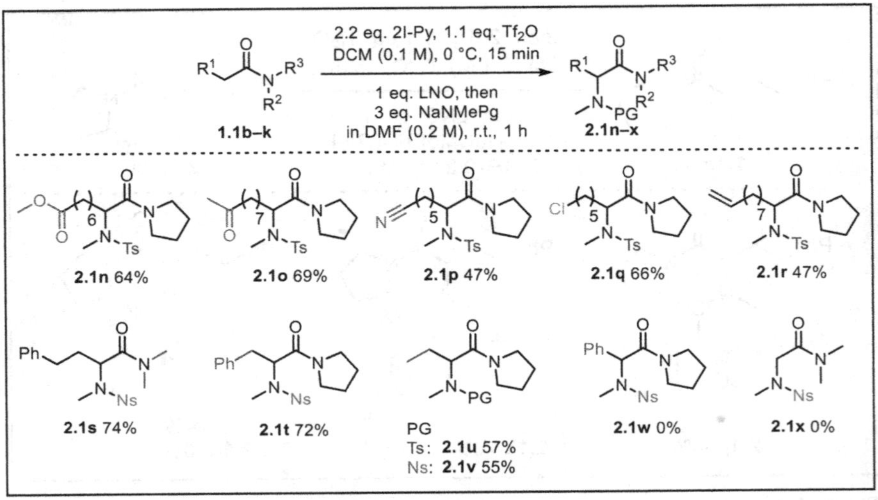

Figure 17: Amide scope of α-amination *via* Umpolung.

Our group has previously shown that the installation of a chiral auxiliary at the amide can induce a good diasterioselectivity in this Umpolung strategy.[27] For this, (*R*)-1-(2-benzhydrylpyrrolidin-1-yl) was chosen as a chiral auxilary. The reaction with methyl tosyl amide led to product **2.1y** in low yield and a poor d.r. (Figure 18). Both diasteromers were isolated as an inseparable mixture. The low yield is probably due to the steric hindrance exerted by the chiral auxiliary.

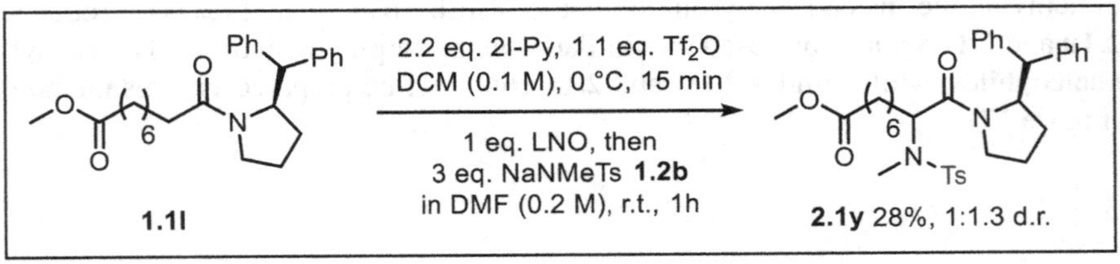

Figure 18: An attempt for a diasteroselective α-amination.

Among the sulfonamide family, the nosyl protecting group is popular because of its reliable deprotection method.[38] A thiol, such as thiophenol, can be used in the combination with a base to trigger a nucleophilic aromatic substitution with irreversible loss of SO_2. This leads to the side product 4-nitrobenzenethiol as well as the free amine. Employing this convenient method, amide **2.1d** was deprotected to form the free α-amino amide **2.1z** in an excellent 93% yield (Figure 19). This detives a two-step sequence whereby an amide is converted to an unsubstituted α-amino amide in an overall yield of 64%.

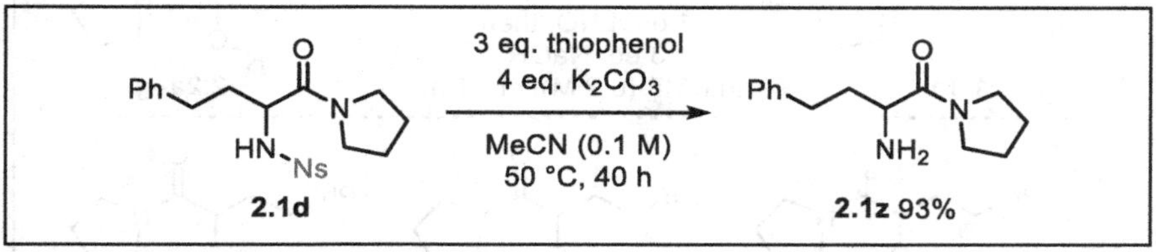

Figure 19: Deprotection of the nosyl group. (Adapted from Gonçalves et al. 2019, https://pubs.acs.org/doi/10.1021/jacs.9b06956; with kind permission from © ACS Publications 2020. All Rights Reserved)

2.1.2   Oxygenation

Results discussed in this chapter include some of Carlos Rafael Gonçalves [those examples will be marked with the prefix **CG** and no experimental data will be provided].

Given the success of sulfonamides in the chemoselective α-amination of amides, we were eager to probe alcohols as a potential nucleophile class for addition to the enolonium ion. In a first experiment, *para*-nitro phenol already gave a good 79% yield (**2.2a**). Benzyl (**CG2.2b**) and allyl (**CG2.2c**) alcohols worked well in this transformation. Surprisingly, ethyl 2-oxocyclohexane-1-carboxylate acted as an oxygen-centered nucleophile and formed compound **CG2.2d**. In a previous report, our group has shown that other β-keto esters react at the α-carbon.[30] Primary metabolite derivatives such as a protected furanose or protected threonine gave products **CG2.2f** and **CG2.2g** respectively in high yields (Figure 20).

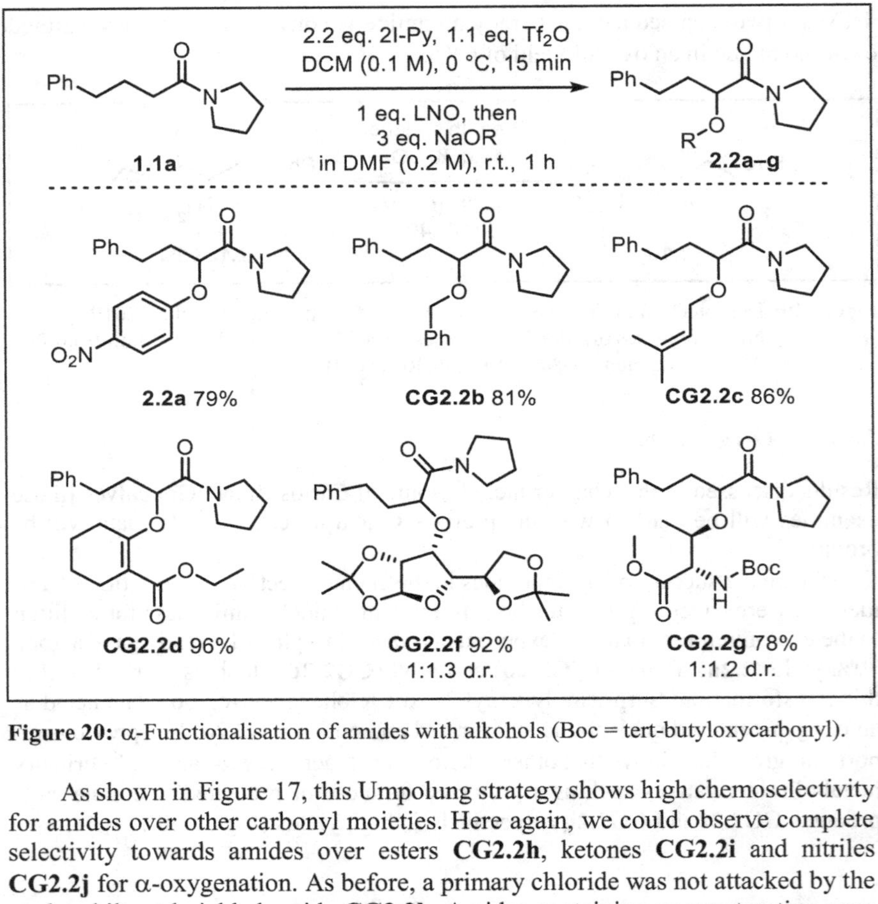

Figure 20: α-Functionalisation of amides with alkohols (Boc = tert-butyloxycarbonyl).

As shown in Figure 17, this Umpolung strategy shows high chemoselectivity for amides over other carbonyl moieties. Here again, we could observe complete selectivity towards amides over esters **CG2.2h**, ketones **CG2.2i** and nitriles **CG2.2j** for α-oxygenation. As before, a primary chloride was not attacked by the nucleophile and yielded amide **CG2.2k**. Amides containing an unsaturation gave similar yields shown with alkene **CG2.2l** and alkyne **CG2.2m** (Figure 21).

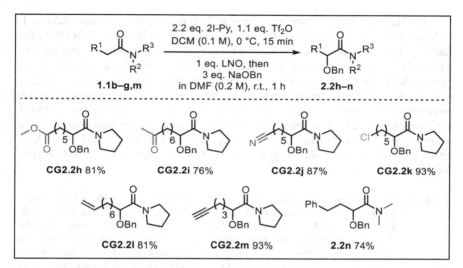

Figure 21: Nucleophilic α-oxygenation of different amides.

This method complements our group's reported α-oxidation of amides.[29]

In the course of our study, we noticed that even weakly nucleophilic DMF could add to the enolonium species. If no other nucleophile, is present a suspension of NaH in DMF reacted with the enolonium ion to form intermediate **A** which upon aqueous work up gave formate **2.2o** in 26% yield and alcohol **2.2p** in 42% (Figure 22). The same reaction carried out with dimethylacetamide instead of DMF did not produce the α-acetylated compound.

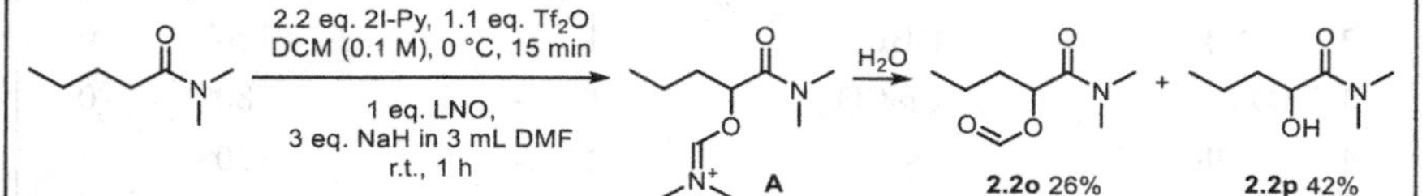

Figure 22: Attack of DMF on the enolonium intermediate.

2.2 Arylation *via* a Smiles rearrangement

With a range of α-sulfonamido amides in hand we were eager to try a Smiles rearrangement as described above in chapter 1.3. This would lead to deprotected and densely functionalised α-amino-α-aryl amides.

For the Smiles rearrangement of amide **2.1b** leading to compound **2.3a** multiple conditions were investigated which are summarised in Table 2. In DMF, the rearrangement only took place when NaH was employed at room temperature (entries **1–7**). In dioxane however, the reaction proceeded with tetrapropylammonium hydroxide (Pr₄NOH) in 42% isolated yield (entry **8**). The yield improved in acetonitrile as a solvent (entry **9**). These conditions were superior to any other which were tested. All other solvents and solvent mixtures investigated gave a lower yield (entries **10–21**). The base Pr₄NOH also proved to be the best, among those examined for this transformation (entries **22–30**). Different additives such as phase transfer catalysts, similar in nature to Pr₄NOH, were less efficient (entries **30–35**).

Table 2:   Optimisation conditions for the Smiles rearrangement of **2.1b** to **2.3a**. (iso.) = isolated yield, a) average yield of two reactions (56% and 62%), DBU = 1,8-Diazabicyclo[5.4.0]undec-7-ene, TBAF = tetrabutylammonium fluoride, TBAB = tetrabutylammonium bromide, TEBnAC = triethylbenzylammonium chloride, 18-crown-6 = 1,4,7,10,13,16-hexaoxacyclooctadecane. NMR yields were determined with 1,3,5-trimethoxybenzene as an internal standard.

en-try	solvent	base	base eq.	additive	T (°C)	NMR Yield
1	DMF	DBU	1	-	80	0
2	DMF	DBU	4	-	120	0
3	DMF	Cs₂CO₃	1	-	80	0
4	DMF	Cs₂CO₃	2	-	120	0
5	DMF	Pr₄NOH	1	-	80	0
6	DMF	NaH	1	-	0	0
7	DMF	NaH	2	-	r.t.	30
8	dioxane	Pr₄NOH	2	-	r.t.	42 (iso.)
9	MeCN	Pr₄NOH	2	-	r.t.	59[a]
10	DCM	Pr₄NOH	2	-	r.t.	4
11	toluene	Pr₄NOH	2	-	r.t.	0

en-try	solvent	base	base eq.	additive	T (°C)	NMR Yield
12	H2O	Pr4NOH	2	-	r.t.	0
13	dioxane/DMF (1/1)	Pr4NOH	2	-	r.t.	44
14	MeCN/DMF (1/1)	Pr4NOH	2	-	r.t.	34
15	DCM/MeCN (1/1)	Pr4NOH	2	-	r.t.	34
16	H2O/MeCN (1/9)	Pr4NOH	2	-	r.t.	0
17	H2O/MeCN (1/9)	DBU	2	-	r.t.	0
18	MeCN (0.5 M)	Pr4NOH	2	-	r.t.	50
19	MeCN (0.01 M)	Pr4NOH	2	-	r.t.	20
20	MeCN	Pr4NOH	2	-	r.t. (1 h)	35
21	MeCN	Pr4NOH	2	-	0	53
22	MeCN	Me4NOH	5	-	r.t.	22
23	MeCN	Bu4NOH	5	-	r.t.	42
24	MeCN	Bu4NOH	2	-	r.t.	45
25	MeCN	NaNH2	5	-	r.t.	8
26	MeCN	NEt3	2	-	r.t.	0
27	MeCN	LHMDS	2	-	r.t.	0
28	MeCN	KHMDS	2	-	r.t.	0
29	MeCN	Proton-sponge	2	-	r.t.	0
30	MeCN	TBAF	2	-	r.t.	0
31	MeCN	KOH	2	TBAB	r.t.	7
32	MeCN	KOH (50% aq.)	20	TBAB	r.t.	29
33	MeCN	KOH (50% aq.)	20	TEBnAC	r.t.	44
34	MeCN	K2CO3	2	TBAB	r.t.	0
35	MeCN	K2CO3	5	18-crown-6	r.t.	0

With the conditions optimised to an acceptable yield, different substrates were tested. Phenylalanine derivative **2.1t** formed the target compound **2.3b** (Figure 24) in 45% yield. The sterically less demanding amide **2.1s** gave product **2.3c** in a slight improved yield of 51% compared to the similar compound **2.3a**. When the *para*-nitro group was changed to an *ortho*-nitro group, the starting amide **2.1j** did not rearrange to **2.3d**. This method was also not suitable for the rearrangement of a pyridine group from **2.1l** to **2.3e** (Figure 23). Because of the high interest of functionalised pyridines, additional conditions were tested for this special case. At 80 °C, **2.1l** did not undergo rearrangement and with 1,8-Diazabicyclo[5.4.0]un-dec-7-ene (DBU) in DMF at room temperature or at 80 °C no reaction took place.

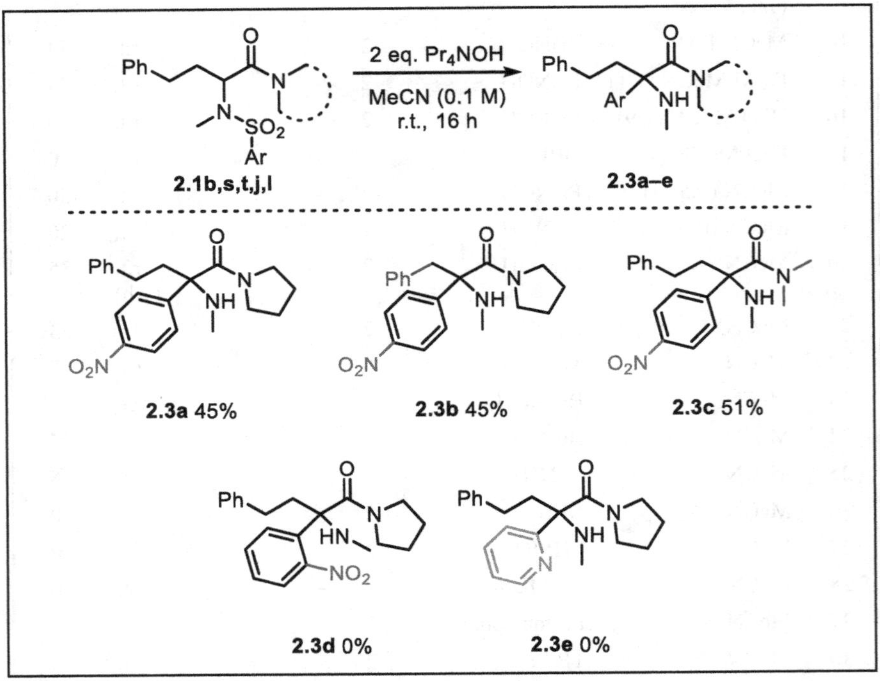

Figure 23: Scope of the Smiles rearrangement on amides.

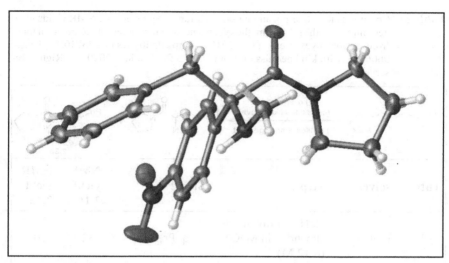

Figure 24: X-ray structure of Smiles product **2.3b**. (Adapted from Gonçalves et al. 2019, https://pubs.acs.org/doi/10.1021/jacs.9b06956; with kind permission from © ACS Publications 2020. All Rights Reserved)

The thought of combining the α-amination and the rearrangement event in a single step was appealing. Despite the moderate yield in the optimisation of the Smiles rearrangement, a few one-pot experiments were carried out. Their results are summarised in Table 3. Entries **2** and **3** were the only cases in which Smiles product **2.3a** was obtained. Huge excess of Pr₄NOH as well as the absence of DMF seemed to be crucial.

Table 3: Conditions for a one-pot amination – rearrangement reaction. NMR yields were determined with 1,3,5-trimethoxybenzene as an internal standard. (Adapted from Gonçalves et al. 2019, https://pubs.acs.org/doi/10.1021/jacs. 9b06956; with kind permission from © ACS Publications 2020. All Rights Reserved)

entry	solvent	Step 1	Step 2	NMR yield 2.1b	NMR yield 2.3a
1	MeCN	NaH + sulfonamide supended in MeCN (0.12 M)	3 eq. Pr₄NOH	32	0
2	MeCN	Na-sulfonamidate	10 eq. Pr₄NOH	4	19
3	MeCN	Na-sulfonamidate + 2 eq. TBAHFP	10 eq. Pr₄NOH	1	15
4	DCM	NaH + sulfonamide in DMF (0.6 M)	Addition of 3 mL MeCN, 10 eq. Bu₄NOH	70	0
5	DCM	NaH + sulfonamide in DMF (0.6 M)	Addition of 3 mL dioxane, 10 eq. Bu₄NOH	69	0

3   Conclusion and Perspectives

3.1   Conclusion

In this study, a novel concept for the α-functionalisation of amides was further developed. The Umpolung of amides enables a nucleophilic α-amination with several sulfonamide derivatives. This method reliably affords good yields and shows a high functional group tolerance. Additionally, α-oxygenation was achieved using similar conditions.

The products were further derivatised through an atom-economical Smiles rearrangement to yield densely functionalised amino acid derivatives.

3.2   Enantioselective α-Amination

The amide Umpolung method developed over the last two years is now a well-established concept. One challenge that still needs to be addressed is enantioselectivity. All of the enolonium-mediated transformations described in this work yield racemic mixtures of products. It has been previously shown that a chiral enolonium intermediate can be generated *via* keteniminium formation by protonation of an ynamide. Using an axially chiral quinoline *N*-oxide derivative, Shin *et al.* observed an *ee* of 83% for the addition of *N*-methyl indole (Figure 25).[39]

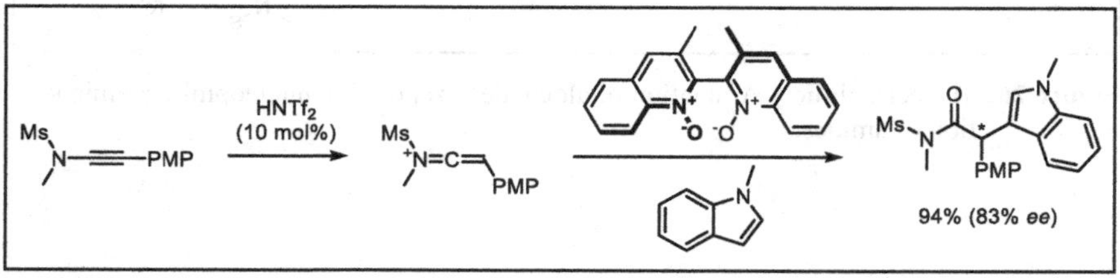

Figure 25: A chiral *N*-oxide promotes enantioselective Umpolung, Ms = methylsulfonyl, PMP = para-methoxyphenyl.

A different approach would be the employment of a chiral counter ion.[40] For the activation of amides, a mixed anhydride might be used to achieve this purpose.

Both of these strategies – a chiral *N*-oxide or a chiral counter ion – could be tested for the enantioselective α-amination of amides.

© The Editor(s) (if applicable) and The Author(s), under exclusive license to Springer Fachmedien Wiesbaden GmbH, part of Springer Nature 2020
M. Lemmerer, *Chemoselective Nucleophilic α-Amination of Amides*, BestMaters, https://doi.org/10.1007/978-3-658-30020-3_3

3.3 γ-Amination

In 2006 Jørgensen *et al.* have reported the first enantioselective γ-amination of aldehydes using a dienamine intermediate. This nucleophilic intermediate reacted with diethyl azodicarboxylate (DEAD) to form γ-aminated products in moderate yield and high *ee* (Figure 26a).[41]

   With this in mind, one can envisage a nucleophilic γ-amination of amides using the chemistry developed in this book. From an α,β-unsaturated amide, a β,γ-unsaturated enolonium might be formed. This electrophilic species could then be attacked by a sulfonamidate to form the y-aminated amide chemoselectively (Figure 26b). One possible challenge that could arise is the competition between a 1,3-attack and the desired 1,5-attack.

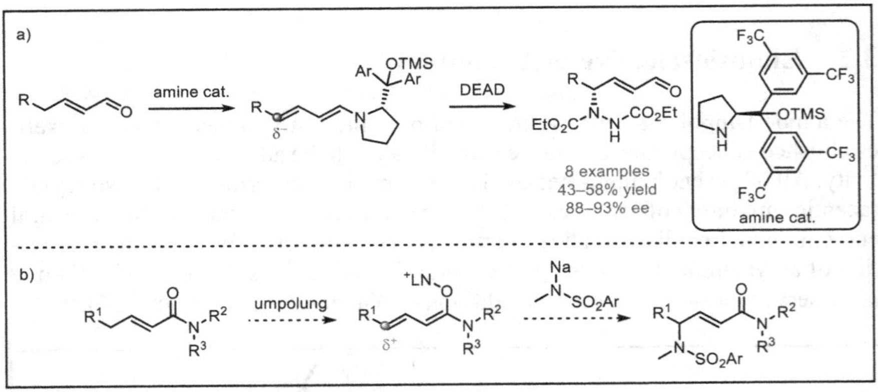

Figure 26: a) electrophilic γ-Amination of aldehydes, b) possible nucleophilic γ-amination of amides.

4 Experimental Data

All glassware was flame-dried before use and the reactions were performed under an atmosphere of argon. All dry solvents were bought from Acros and used as received. All other reagents, the synthesis of which is not described below, were used as received from commercial suppliers. Reaction progress was monitored by thin layer chromatography (TLC) performed on aluminum plates coated with silica gel F254 with 0.2 mm thickness. Chromatograms were visualized by fluorescence quenching with UV light at 254 nm or by staining using potassium permanganate. Column chromatography was performed using silica gel 60 (230-400 mesh, Merck and co.). Neat infra-red spectra were recorded using a Perkin-Elmer Spectrum 100 FT-IR spectrometer. Wavenumbers ($= 1/\lambda$) are reported in cm^{-1}. Mass spectra were obtained using a Finnigan MAT 8200 or (70 eV) or an Agilent 5973 (70 eV) spectrometer, using electrospray ionisation (ESI). All ^{1}H NMR and ^{13}C NMR spectra were recorded using a Bruker AV-400, AV-600 or AV-700 spectrometer at 300K. Chemical shifts were given in parts per million (ppm, δ), referenced to the solvent peak of CDCl$_3$, defined at $\delta = 7.26$ ppm (^{1}H NMR) and $\delta = 77.16$ (^{13}C NMR). Coupling constants are quoted in Hz (J). ^{1}H NMR splitting patterns were designated as singlet (s), doublet (d), triplet (t), quartet (q). Splitting patterns that could not be interpreted or easily visualized were designated as multiplet (m).

Amides

Amides **1.1b,d,e,h–m** were either generously donated by other members of our group or received from commercial suppliers.

General procedure A for the synthesis of amides

To the a solution of the acid (1 eq.) in DCM (0.2 M) was added 1-hydroxybenzotriazole hydrate (1 eq.), 1-(3-dimethylaminopropyl)-3-ethylcarbodiimide hydrochloride (1 eq.), triethylamine (1 eq.) and the secondary amine (1 eq.). The solution was stirred for 16 h. After dilution in AcOEt, the organic phase was washed subsequently with a solution of HCl (1 M), saturated solution of NaHCO$_3$ and finally with a saturated solution of NaCl. The organic layer was separated, dried over MgSO$_4$ and the solvent was removed under reduced pressure. The crude mixture was purified by column chromatography on silica gel.

© The Editor(s) (if applicable) and The Author(s), under exclusive license to Springer Fachmedien Wiesbaden GmbH, part of Springer Nature 2020
M. Lemmerer, *Chemoselective Nucleophilic α-Amination of Amides*, BestMaters, https://doi.org/10.1007/978-3-658-30020-3_4

1.1a 4-Phenyl-1-(pyrrolidin-1-yl)butan-1-one

The product was prepared from 2.46 g (15 mmol) 4-phenylbutyric acid and 1.07 g (15 mmol) pyrrolidine according to general procedure A. Purification by column chromatography (EtOAc:heptane = 2:1) gave the amide in 93% (3.04 g). All spectroscopic properties are in good accordance with reported data.[23]

1.1c 1-(Pyrrolidin-1-yl)undecane-1,10-dione

This amide was synthesised according to a reported procedure in 69% yield.[25] All spectroscopic properties are in good accordance with reported data.[25]

1.1f 1-(Pyrrolidin-1-yl)undec-10-en-1-one

To a solution of 711 mg (10 mmol) pyrrolidine in 33 mL DCM (0.15 M) was added 2.02 g (20 mmol) triethylamine and 2.43 g (12 mmol) 10-undecenoyl chloride and the solution was stirred for 16 h. Then the reaction was quenched with a saturated solution of NH$_4$Cl and extracted with EtOAc. The organic layer was separated and the solvent removed under reduced pressure. The crude mixture was purified by column chromatography on silica gel (EtOAc:heptane = 1:1) to yield amide **1.1f** in 92% (2.19 g). All spectroscopic properties are in good accordance with reported data.[42]

1.1g *N,N*-Dimethyl-4-phenylbutanamide

The product was prepared from 2.46 g (15 mmol) 4-phenylbutyric Acid and 7.5 mL (15 mmol) dimethylamine (2 M in THF) according to general procedure A. Purification by column chromatography on silica gel (EtOAc:heptane = 2:1) afforded amide **1.1g** in 71% (2.04 g) yield. All spectroscopic properties are in good accordance with reported data.[43]

Sulfonamides

Sulfonamides **1.2b,c,g** were either generously donated by other members of our group or received from commercial suppliers.

General procedure B for the synthesis of sulfonamides

To a solution of sulfonyl chloride (1 eq.) in THF (1 M) was added the primary amine (3 eq.) or the primary amine hydrochloride salt (1 eq.) in combination with triethylamine (2.5 eq.) at room temperature. The mixture was stirred for 16 h while a precipitate formed. Then a solution of HCl (1 M) was added until a pH of 2 was reached. The solution was extracted with EtOAc. The combined organic layers were washed with a saturated solution of NaCl and dried over $MgSO_4$. The solvent was removed under reduced pressure which yielded the product. The crude product was used without further purification.

1.2a *N*-Methyl-4-nitrobenzenesulfonamide

The product was prepared from 2.22 g (10 mmol) 4-nitrobenzenesulfonyl chloride and 15 mL (30 mmol) methylamine (2 M in THF) in 96% (2.12 g) yield according to the general procedure B. All spectroscopic properties are in good accordance with reported data.[44]

1.2d *N*-Isopropyl-4-methylbenzenesulfonamide

The product was prepared from 381 mg (2 mmol) 4-methylbenzenesulfonyl chloride and 355 mg (6 mmol) *iso*-propylamine in 95% (404 mg) yield according to the general procedure B. All spectroscopic properties are in good accordance with reported data.[45]

1.2e *N*-Benzyl-4-methylbenzenesulfonamide

The product was prepared from 381 mg (2 mmol) 4-methylbenzenesulfonyl chloride and 643 mg (6 mmol) benzylamine in 94% (489 mg) yield according to the general procedure B. All spectroscopic properties are in good accordance with reported data.[46]

1.2f 4-Methyl-*N*-phenylbenzenesulfonamide

The product was prepared from 381 mg (2 mmol) 4-methylbenzenesulfonyl chloride and 559 mg (6 mmol) aniline in 87% (432 mg) yield according to the general procedure B. All spectroscopic properties are in good accordance with reported data.[46]

1.2h Methyl tosylglycinate

The product was prepared from 381 mg (2 mmol) 4-methylbenzenesulfonyl chloride, 281 mg (2 mmol) glycine methyl ester hydrochloride and 506 mg (5 mmol) trimethylamine in 91% (445 mg) yield according to the general procedure B. All spectroscopic properties are in good accordance with reported data.[47]

1.2i *N*-Methyl-2-nitrobenzenesulfonamide

The product was prepared from 665 mg (3 mmol) 2-nitrobenzenesulfonyl chloride and 4.5 mL (9 mmol) methylamine (in THF 2 M) in 93% (600 mg) yield according to the general procedure B. All spectroscopic properties are in good accordance with reported data.[48]

1.2j *N*-Methylnaphthalene-2-sulfonamide

The product was prepared from 453 mg (2 mmol) naphthalene-2-sulfonyl chloride and 3 mL (6 mmol) methylamine (in THF 2 M) in 96% (423 mg) yield according to the general procedure B. All spectroscopic properties are in good accordance with reported data.[49]

The synthesis of pyridine derived sulfonamides **1.2l,m** is a modified version of Wright *et al.*'s method.[50]

1.2k *N*-Methylpyridine-2-sulfonamide

2-Mercaptopyridine (0.561 g, 5 mmol) was stirred in a mixture of 25 mL of DCM and 25 mL of a 1 M solution of HCl for 10 min at -10 to -5 °C. Then, a cold (5 °C) solution of sodium hypochlorite (11% (aq.), 9.3 mL, 16.5 mmol) was added dropwise under vigorous stirring. After the addition was completed, the mixture was stirred for 15 min at -10 to -5 °C. It was transferred to a separatory funnel (pre-cooled in the freezer) and the organic layer was rapidly separated and collected in an Erlenmeyer flask cooled in a dry ice-acetone bath. Methylamine (2 M in THF, 6.25 mL, 12.5 mmol) was added while stirring, whereupon the organic layer became a white suspension. The suspension was stirred for 30 min at 0 °C. Then, the mixture was washed with 1 M solution of HCl, water and a solution of saturated NaCl, dried with MgSO$_4$ and concentrated under reduced pressure. Purification by column chromatography on silica gel (EtOAc:heptane = 1:1) yielded the product as a white solid (400 mg, 47%).[50]

^{1}H NMR (400 MHz, CDCl$_3$): δ = 8.75–8.69 (m, 1H^8), 8.02 (dt, *J* = 7.7, 0.9 Hz, 1H^{11}), 7.93 (td, *J* = 7.7, 1.7 Hz, 1H^{10}), 7.51 (ddd, *J* = 7.6, 4.7, 1.2 Hz, 1H^9), 4.92 (s, 1H^{NH}), 2.75 (d, *J* = 5.3 Hz, 3H^6) ppm; **^{13}C NMR (175 MHz, CDCl$_3$)**: δ = 156.7 (C^2), 150.2 (CH^{Ar}), 138.2 (CH^{Ar}), 126.8 (CH^{Ar}), 122.7 (CH^{Ar}), 23.0 (CH^6) ppm; **IR (neat)**: $\overline{V}$ = 3289, 1579, 1454, 1428, 1325, 1174, 1117, 1086, 992, 844, 777, 738, 590 cm^{-1}; **HRMS (ESI)**: m/z calculated for [M+Na]$^+$ (C$_6$H$_8$N$_2$O$_2$SNa) 195.0199, found 195.0198.

1.2l *N*-Methyl-5-nitropyridine-2-sulfonamide

5-Nitro-2-mercaptopyridine (312 mg, 2 mmol) was stirred in a mixture of 10 mL of DCM and 10 mL of a 1 M solution of HCl for 10 min at 0 °C. Then, a cold (0 °C) solution of sodium hypochlorite (11% (aq.), 26 mL, 6.6 mmol) was added dropwise under vigorous stirring. After the addition was completed, the mixture was stirred for 60 min at 0 °C. It was transferred to a separatory funnel (pre-cooled in the freezer) and the organic layer was rapidly separated and collected in an Erlenmeyer flask cooled in a dry ice-acetone bath. Methylamine (2 M in THF, 2.5 mL, 5 mmol) was added while stirring, whereupon the organic layer became a white suspension. The suspension was stirred for 30 min at 0 °C. Then, the mixture was washed with a 1 M solution of HCl, water and a solution of saturated NaCl, dried with $MgSO_4$ and concentrated under reduced pressure. Purification by column chromatography on silica gel (EtOAc:heptane = 1:20–1:4) yielded the product as a white solid (92 mg, 21%).[50]

^{1}H NMR (400 MHz, CDCl$_3$): δ = 9.49 (dd, , J = 2.5, 0.5 Hz, 1H^8), 8.71 (dd, J = 8.5, 2.5 Hz, 1H^{10}), 8.22 (dd, J = 8.5, 0.5 Hz, 1H^{11}), 4.86 (bs, 1H^{NH}), 2.85 (d, J = 5.2 Hz, 3H^6) ppm; **^{13}C NMR (175 MHz, CDCl$_3$):** δ = 161.8 (C^{Ar}), 145.6 (CH^{Ar}), 145.4 (C^{Ar}), 133.6 (CH^{Ar}), 123.0 (CH^{Ar}), 30.2 (CH^6) ppm; **IR (neat):** $\overline{\nu}$ = 3283, 3099, 1600, 1567, 1531, 1355, 1334, 1176, 1104, 1016, 840, 750, 612 cm^{-1}; **HRMS (ESI):** m/z calculated for [M+Na]$^+$ (C$_6$H$_7$N$_3$O$_4$SNa) 240.0045, found 240.0049.

General procedure C for the synthesis of α-functionalised amides 2.1a–y and 2.2a,n

To a solution of amide (0.2 mmol) in DCM (2 mL) was added 2-iodopyridine (46.8 μL, 0.44 mmol). The reaction mixture was cooled to 0 °C before triflic anhydride (37 μL, 0.22 mmol) was added. After stirring for 15 min, lutidine *N*-oxide (22.4 μL, 0.2 mmol) was added and the reaction was stirred for a further 5 min at 0 °C. Then, a solution of the sodium amide generated from addition of the amine, amide or alcohol (0.6 mmol) to a suspension of NaH (0.6 mmol) in DMF (3 mL) was added. The reaction mixture was stirred at room temperature for 1 h before being quenched with a saturated solution of NH$_4$Cl. The layers were separated and the aqueous layer was extracted with DCM. The combined organic layers were washed with brine before being dried over $MgSO_4$. The solvent was removed under reduced pressure.

2.1a 2-(1*H*-Indol-1-yl)-4-phenyl-1-(pyrrolidin-1-yl)butan-1-one

The product was prepared according to general procedure C. Purification by column chromatography on silica gel (EtOAc:heptane = 1:2) yielded the product (51.4 mg, 77%) as a pale yellow liquid.

^{1}H NMR (400 MHz, CDCl$_3$): δ = 7.63 (d, J = 7.7 Hz, 1H^{Ar}), 7.34–7.08 (m, 9H^{Ar}), 6.55 (d, J = 3.3 Hz, 1H^{13}), 4.97–4.92 (m, 1H^1), 3.58–3.39 (m, 2H^6), 3.28–3.20 (m, 1H^9), 2.88–2.79 (m, 1H^9), 2.65–2.52 (m, 3H^{Alk}), 2.43–2.31 (m, 1H^{Alk}), 1.87–1.66 (m, 4H^{Alk}), 1.51–1.40 (m, 1H^{Alk}) ppm; **^{13}C NMR (175 MHz, CDCl$_3$):** δ = 167.8 (C^2), 140.8 (C^{Ar}), 136.1 (C^{Ar}), 128.7 (C^{Ar}), 128.6 (2CH^{Ar}), 128.6 (2CH^{Ar}), 126.3 (CH^{Ar}), 126.2 (CH^{Ar}), 121.8 (CH^{Ar}), 121.2 (CH^{Ar}), 119.8 (CH^{Ar}), 109.1 (CH^{Ar}), 102.3 (CH^{Ar}), 56.2 (CH^1), 46.4 (CH$_2^6$), 46.2 (CH$_2^9$), 33.8 (CH$_2^{Alk}$), 31.9 (CH$_2^{Alk}$), 26.2 (CH$_2^{Alk}$), 24.0 (CH$_2^{Alk}$) ppm; **IR (neat):** $\tilde{V}$ = 3026, 2969, 2875, 1645, 1456, 1400, 1308, 1192, 700 cm^{-1}; **HRMS (ESI):** m/z calculated for [M+Na]$^+$ (C$_{22}$H$_{24}$N$_2$ONa) 355.1781, found 355.1780.

2.1b *N*-Methyl-4-nitro-*N*-(1-oxo-4-phenyl-1-(pyrrolidin-1-yl)butan-2-yl)benzenesulfonamide

The product was prepared according to general procedure C. Purification by column chromatography on silica gel (EtOAc:heptane = 1:2) yielded the product (82.3 mg, 82%) as a pale yellow liquid.

^{1}H NMR (400 MHz, CDCl$_3$): δ = 8.30 (d, J = 9.0 Hz, 2H^{25,27}), 7.82 (d, J = 9.0 Hz, 2H^{24,28}), 7.35–7.29 (m, 2H^{16,18}), 7.27–7.21 (m, 1H^{17}), 7.15–7.10 (m, 2H^{15,19}), 4.63 (apt, J = 7.5 Hz, 1H^1), 3.64–3.57 (m, 1H^6), 3.41–3.19 (m, 3H^{6,9}), 3.02 (s, 3H^{11}), 2.67–2.50 (m, 2H^{13}), 2.18–2.07 (m, 1H^{Alk}), 1.98–1.80 (m, 4H^{Alk}), 1.65–1.55 (m, 1H^{Alk}) ppm; **^{13}C NMR (100 MHz, CDCl$_3$):** δ = 167.7 (C^2), 150.1 (C^{26}), 144.8 (C^{Ar}), 140.4 (C^{Ar}), 128.8 (2CH^{Ar}), 128.6 (2CH^{Ar}), 128.5 (2CH^{Ar}), 126.7 (CH^{17}), 124.3 (2CH^{Ar}), 56.4 (CH^1), 46.5 (CH$_2^6$), 46.1 (CH$_2^6$), 32.1(CH$_2^{Alk}$), 30.8 (CH$_3^{11}$), 30.3 (CH$_2^{Alk}$), 26.3 (CH$_2^{Alk}$), 24.2 (CH$_2^{Alk}$) ppm; **IR (neat):** $\tilde{V}$ = 3103, 3063, 2974, 2876, 1642, 1529, 1448, 1349, 1162, 855, 699, 608 cm^{-1}; **HRMS (ESI):** m/z calculated for [M+Na]$^+$ (C$_{21}$H$_{25}$N$_3$O$_5$SNa) 454.1407, found 454.1411.

2.1c *N*,4-Dimethyl-*N*-(1-oxo-4-phenyl-1-(pyrrolidin-1-yl)butan-2-yl)benzenesul-fonamide

The product was prepared according to general procedure C. Purification by column chromatography on silica gel (EtOAc:heptane = 1:2) yielded the product (57.0 mg, 71%) as a pale yellow liquid.

^{1}H NMR (400 MHz, CDCl$_3$): δ = 7.58 (d, *J* = 7.8 Hz, 2H^{24,28}), 7.32–7.18 (m, 5H^{Ar}), 7.09 (d, *J* = 7.8 Hz, 2H^{Ar}), 4.59 (dd, *J* = 9.8, 6.4 Hz, 1H^1), 3.76–3.68 (m, 1H^6), 3.46–3.21 (m, 3H^{6,9}), 2.92 (s, 3H^{11}), 2.64–2.54 (m, 1H^{Alk}), 2.50–2.41 (m, 1H^{Alk}), 2.42 (s, 3H^{29}), 2.18–2.09 (m, 1H^{Alk}), 1.98–1.78 (m, 4H^{Alk}), 1.51–1.40 (m, 1H^{Alk}) ppm; **^{13}C NMR (100 MHz, CDCl$_3$)**: δ = 167.7 (C^2), 143.6 (C^{Ar}), 140.9 (C^{Ar}), 136.2 (C^{Ar}), 129.8 (2CH^{Ar}), 128.6 (2CH^{Ar}), 128.6 (2CH^{Ar}), 127.4 (2CH^{Ar}), 126.4 (CH^{17}), 56.5 (CH^1), 46.6 (CH$_2^6$), 46.1 (CH$_2^9$), 32.4 (CH$_2^{Alk}$), 30.5 (CH$_3^{11}$), 29.6 (CH$_2^{Alk}$), 26.3 (CH$_2^{Alk}$), 24.3 (CH$_2^{Alk}$), 21.7 (CH$_3^{29}$) ppm; **IR (neat)**: $\bar{V}$ = 3059, 3026, 2950, 2874, 1740, 1639, 1441, 1335, 1156, 1087, 933, 814, 696, 651 cm^{-1}; **HRMS (ESI)**: m/z calculated for [M+Na]$^+$ (C$_{22}$H$_{28}$N$_2$O$_3$SNa) 423.1713, found 423.1716.

2.1d 4-Nitro-*N*-(1-oxo-4-phenyl-1-(pyrrolidin-1-yl)butan-2-yl)benzenesulfona-mide

The product was prepared according to general procedure C. Purification by column chromatography on silica gel (EtOAc:heptane = 1:2) yielded the product (57.2 mg, 69%) as a white solid.

^{1}H NMR (400 MHz, CDCl$_3$): δ = 8.29 (d, J = 8.8 Hz, 2H^{25,27}), 7.99 (d, J = 8.8 Hz, 2H^{24,28}), 7.33–7.16 (m, 5H^{Ar}), 5.86 (bs, 1H^N), 3.94 (bs, 1H^1), 3.20–3.11 (m, 1H^6), 2.99–2.88 (m, 3H^{6,9}), 2.87–2.68 (m, 2H^{13}), 1.94–1.58 (m, 6H^{Alk}) ppm; **^{13}C NMR (100 MHz, CDCl$_3$):** δ = 168.5 (C^2), 150.2 (C^{Ar}), 145.8 (C^{Ar}), 140.5 (C^{Ar}), 128.8 (CH^{Ar}), 128.7 (CH^{Ar}), 126.6 (2CH^{17}), 124.1 (2CH^{Ar}), 54.1 (CH^1), 46.1 (CH$_2$^{6}), 45.9 (CH$_2$^{9}), 35.1 (CH$_2$^{Alk}), 31.3 (CH$_2$^{Alk}), 26.0 (CH$_2$^{Alk}), 24.0 (CH$_2$^{Alk}) ppm; **IR (neat):** $\overline{V}$ = 3104, 2952, 2876, 1629, 1528, 1451, 1348, 1164, 1091, 855, 613 cm^{-1}; **HRMS (ESI):** m/z calculated for [M+Na]$^+$ (C$_{20}$H$_{23}$N$_3$O$_5$SNa) 440.1251, found 440.1251.

2.1e   *N*-Isopropyl-4-methyl-*N*-(1-oxo-4-phenyl-1-(pyrrolidin-1-yl)butan-2-yl)-benzenesulfonamide

The product was prepared according to general procedure C. Purification by column chromatography on silica gel (EtOAc:heptane = 1:3) yielded the product (54.6 mg, 64%) as a pale yellow solid.

^{1}H NMR (400 MHz, CDCl$_3$): δ = 7.60 (d, J = 7.8 Hz, 2H^{24,28}), 7.31–7.17 (m, 5H^{Ar}), 7.04 (d, J = 7.8 Hz, 2H^{25,27}), 4.40 (dd, J = 10.3, 4.1 Hz, 1H^1), 4.15–4.05 (m, 1H^{11}), 3.72–3.64 (m, 1H^6), 3.56–3.38 (m, 2H^9), 3.24–3.16 (m, 1H^6), 2.65–2.56 (m, 1H^{Alk}), 2.42 (s, 3H^{29}), 2.41–2.24 (m, 2H^{Alk}), 1.94–1.78 (m, 4H^{Alk}), 1.52–1.42 (m, 1H^{Alk}), 1.33 (d, J = 6.9 Hz, 6H^{23,30}) ppm; **^{13}C NMR (100 MHz, CDCl$_3$):** δ = 167.5 (C^2), 143.1 (C^{Ar}), 141.0 (C^{Ar}), 139.6 (C^{Ar}), 129.6 (2CH^{Ar}), 128.7 (2CH^{Ar}), 128.6 (2CH^{Ar}), 127.6 (2CH^{Ar}), 126.4 (CH^{17}), 57.7 (CH^1), 50.4 (CH^{11}), 46.5 (CH$_2$^{6}), 46.4 (CH$_2$^{9}), 32.7 (CH$_2$^{Alk}), 31.2 (CH$_2$^{Alk}), 26.4 (CH$_2$^{Alk}), 24.2 (CH$_2$^{Alk}), 23.3 (CH$_3$^{23}), 22.4 (CH$_3$^{30}), 21.6 (CH$_3$^{29}) ppm; **IR (neat):** $\overline{V}$ = 3026, 2970, 2873, 1637, 1599, 1439, 1328, 1151, 1085, 965, 936, 815, 698, 597 cm^{-1}; **HRMS (ESI):** m/z calculated for [M+Na]$^+$ (C$_{24}$H$_{32}$N$_2$O$_3$SNa) 451.2026, found 451.2022.

2.1f *N*-Benzyl-4-methyl-*N*-(1-oxo-4-phenyl-1-(pyrrolidin-1-yl)butan-2-yl)benzenesulfonamide

The product was prepared according to general procedure C. Purification by column chromatography on silica gel (EtOAc:heptane = 1:2) yielded the product (67.0 mg, 70%) as a pale yellow solid.

¹H NMR (400 MHz, CDCl₃): δ = 7.62 (d, *J* = 8.1 Hz, 2H²⁸,²⁴), 7.44 (d, *J* = 7.1 Hz, 2H^Ar), 7.32–7.14 (m, 8H^Ar), 7.98–7.93 (m, 2H^Ar), 4.87 (d, *J* = 16.1 Hz, 1H¹¹), 4.26 (dd, *J* = 9.1, 4.8 Hz, 1H¹), 4.33 (d, *J* = 16.1 Hz, 1H¹¹), 3.55–3.46 (m, 1H⁶), 3.20–3.11 (m, 2H⁶,⁹), 2.68–2.59 (m, 1H⁹), 2.55–2.47 (m, 1H¹³), 2.43 (s, 3H²⁹), 2.37–2.27 (m, 1H¹³), 2.24–2.13 (m, 1H³), 1.80–1.61 (m, 3H⁷,⁸), 1.53–1.42 (m, 1H⁸), 1.41–1.31 (m, 1H⁸) ppm; **¹³C NMR (100 MHz, CDCl₃):** δ = 166.7 (C²), 143.7 (C^Ar), 141.1 (C^Ar), 137.9 (C^Ar), 137.7 (C^Ar), 129.8 (CH^Ar), 128.7(CH^Ar), 128.5 (CH^Ar), 128.5 (CH^Ar), 128.2 (CH^Ar), 127.5 (CH^Ar), 127.4 (CH^Ar), 126.3 (CH^Ar), 57.2 (CH¹), 48.6 (CH₂¹¹), 46.1 (CH₂⁶), 45.9 (CH₂⁹), 32.7 (CH₂^Alk), 30.2 (CH₂^Alk), 26.2 (CH₂^Alk), 23.9 (CH₂^Alk), 21.7 (CH₃²⁹) ppm; **IR (neat):** $\tilde{V}$ = 3061, 3027, 2948, 2874, 1642, 1599, 1494, 1435, 1335, 1156, 1093, 698, 654 cm⁻¹; **HRMS (ESI):** m/z calculated for [M+H]⁺ (C₂₈H₃₃N₂O₃S) 477.2206, found 477.2201.

2.1g   4-Methyl-*N*-(1-oxo-4-phenyl-1-(pyrrolidin-1-yl)butan-2-yl)-*N*-phenylben-
zenesulfonamide

The product was prepared according to general procedure C. Purification by column chromatography on silica gel (EtOAc:heptane = 1:3) yielded the product
(69.7 mg, 75%) as a white solid.

¹H NMR (400 MHz, CDCl₃): δ = 7.36 (d, *J* = 8.6 Hz, 2H^{Ar}), 7.32–7.19 (m,
7H^{Ar}), 7.18–7.10 (m, 3H^{Ar}), 7.98–7.93 (m, 2H^{Ar}), 5.04–4.96 (m, 1H^1), 3.88–3.78
(m, 1H^6), 3.40–3.22 (m, 3H^{6,9}), 2.59–2.42 (m, 2H^{13}), 2.34 (s, 3H^{29}), 2.03–1.64 (m,
6H^{Alk}) ppm; **¹³C NMR (100 MHz, CDCl₃):** δ = 167.8 (C^2), 143.5 (C^{Ar}), 140.9
(C^{Ar}), 137.0 (C^{Ar}), 136.0 (C^{Ar}), 132.4 (2CH^{Ar}), 129.3 (2CH^{Ar}), 128.8 (CH^{10}), 128.7
(2CH^{Ar}), 128.7 (2CH^{Ar}), 128.6 (2CH^{Ar}), 127.9 (2CH^{Ar}), 126.4 (CH^{17}), 59.1 (CH^1),
46.4 (CH₂^6), 46.3 (CH₂^9), 32.6 (CH₂^{Alk}), 32.4 (CH₂^{Alk}), 26.5 (CH₂^{Alk}), 24.3
(CH₂^{Alk}), 21.7 (CH₃^{29}) ppm; **IR (neat):** $\widetilde{V}$ = 3061, 3026, 2951, 2874, 2242, 1643,
1597, 1491, 1439, 1342, 1159, 909, 729, 697, 657 cm⁻¹; **HRMS (ESI):** m/z calculated for [M+H]⁺ (C₂₇H₃₁N₂O₃S) 463.2050, found 463.2046.

2.1h   *N*-(4-Methoxyphenyl)-4-methyl-*N*-(1-oxo-4-phenyl-1-(pyrrolidin-1-yl)bu-tan-2-yl)benzenesulfonamide

The product was prepared according to general procedure C. Purification by column chromatography on silica gel (EtOAc:heptane = 1:2) yielded the product (70.4 mg, 72%) as a pale yellow solid.

^{1}H NMR (400 MHz, CDCl$_3$): δ = 7.40 (d, *J* = 8.3 Hz, 2H^{24,28}), 7.31–7.16 (m, 7H^{Ar}), 7.13–7.08 (m, 2H^{Ar}), 6.77 (d, *J* = 9.0 Hz, 2H^{5,12}), 5.03 (dd, *J* = 8.1, 6.4 Hz, 1H^1), 3.93–3.85 (m, 1H^6), 3.78 (s, 3H^{22}), 3.45–3.28 (m, 3H^{6,9}), 2.64–2.48 (m, 2H^{13}), 2.39 (s, 3H^{29}), 2.07–1.82 (m, 6H^{Alk}) ppm; **^{13}C NMR (100 MHz, CDCl$_3$):** δ = 167.9 (C^2), 159.7 (C^{Ar}), 143.4 (C^{Ar}), 141.0 (C^{Ar}), 137.0 (C^{Ar}), 133.5 (2CH^{Ar}), 129.2 (2CH^{Ar}), 128.6 (2CH^{Ar}), 128.6 (2CH^{Ar}), 128.2 (C^{Ar}), 127.9 (2CH^{Ar}), 126.3 (CH^{17}), 113.9 (2CH^{Ar}), 59.0 (CH$_3$^{22}), 55.4 (CH^1), 46.4 (CH$_2$^{6}), 46.2 (CH$_2$^{9}), 32.5 (CH$_2$^{Alk}), 32.4 (CH$_2$^{Alk}), 26.5 (CH$_2$^{Alk}), 24.3 (CH$_2$^{Alk}), 21.6 (CH$_3$^{29}) ppm; **IR (neat):** $\overline{\nu}$ = 3062, 3026, 2952, 2874, 2240, 1644, 1602, 1505, 1439, 1341, 1249, 1158, 914, 727, 589 cm^{-1}; **HRMS (ESI):** m/z calculated for [M+H]$^+$ (C$_{28}$H$_{33}$N$_2$O$_4$S) 493.2156, found 493.2164.

2.1i Methyl *N*-(1-oxo-4-phenyl-1-(pyrrolidin-1-yl)butan-2-yl)-*N*-tosylglycinate

The product was prepared according to general procedure C. Purification by column chromatography on silica gel (EtOAc:heptane = 1:2) yielded the product (58.0 mg, 63%) as a pale yellow liquid.

^{1}H NMR (400 MHz, CDCl$_3$): δ = 7.66 (d, *J* = 8.3 Hz, 2H^{Ar}), 7.32–7.16 (m, 5H^{Ar}), 7.04 (d, *J* = 8.3 Hz, 2H^{Ar}), 4.39 (d, *J* = 18.3 Hz, 1H^{11}), 3.39–3.34 (m, 1H^1), 4.17 (d, *J* = 18.3 Hz, 1H^{11}), 3.71 (s, 3H^{12}), 3.54–3.45 (m, 1H^6), 3.38–3.22 (m, 2H^9), 3.02–2.93 (m, 1H^6), 2.66–2.57 (m, 1H^{Alk}), 2.53–2.44 (m, 1H^{Alk}), 2.43 (s, 3H^{29}), 2.12–2.01 (m, 1H^{Alk}), 1.90–1.74 (m, 4H^{Alk}), 1.62–1.51 (m, 1H^{Alk}) ppm; **^{13}C NMR (100 MHz, CDCl$_3$):** δ = 170.5 (C^4), 167.4 (C^2), 144.0 (C^{Ar}), 140.7 (C^{Ar}), 137.0 (C^{Ar}), 129.8 (2CH^{Ar}), 128.6 (2CH^{Ar}), 128.6 (2CH^{Ar}), 127.8 (2CH^{Ar}), 126.4 (CH^{17}), 55.7 (CH^1), 52.3 (CH^{12}), 46.2 (CH$_2^6$), 46.1 (CH$_2^9$), 45.3 (CH$_2^{11}$), 31.9 (CH$_2^{Alk}$), 30.9 (CH$_2^{Alk}$), 26.2 (CH$_2^{Alk}$), 24.2 (CH$_2^{Alk}$), 21.7 (CH$_3^{29}$) ppm; **IR (neat):** $\overline{V}$ = 2951, 2876, 2251, 1756, 1638, 1598, 1447, 1341, 1206, 1156, 1093, 909, 813, 727, 700, 654 cm^{-1}; **HRMS (ESI):** m/z calculated for [M+Na]$^+$ (C$_{24}$H$_{30}$N$_2$O$_5$SNa) 481.1768, found 481.1758.

2.1j *N*-Methyl-2-nitro-*N*-(1-oxo-4-phenyl-1-(pyrrolidin-1-yl)butan-2-yl)benzene-sulfonamide

The product was prepared according to general procedure C. Purification by column chromatography on silica gel (EtOAc:heptane = 1:4-7:3) yielded the product (69.0 mg, 80%) as a pale yellow solid.

¹H NMR (400 MHz, CDCl₃): δ = 7.94–7.90 (m, 1H^Ar), 7.71–7.60 (m, 3H^Ar), 7.30–7.12 (m, 5H^Ar), 4.75 (apt, *J* = 7.4 Hz, 1H¹), 3.70–3.61 (m, 1H⁶), 3.44–3.37 (m, 2H⁹), 3.31–3.23 (m, 1H⁶), 3.14 (s, 3H¹¹), 2.71–2.56 (m, 2H^Alk), 2.21–2.12 (m, 1H^Alk), 1.95–1.78 (m, 5H^Alk) ppm; **¹³C NMR (100 MHz, CDCl₃):** δ = 168.3 (C²), 148.2 (C²⁴), 140.7 (C^Ar), 133.6 (CH^Ar), 132.6 (C^Ar), 131.7 (CH^Ar), 130.7 (CH^Ar), 128.6 (2CH^Ar), 128.5 (2CH^Ar), 126.3 (CH^Ar), 124.3 (CH^Ar), 56.6 (CH¹), 46.5 (C₂H⁶), 46.1 (C₂H⁹), 32.2 (CH₂^Alk), 31.0 (CH₃¹¹), 30.9 (CH₂^Alk), 26.2 (CH₂^Alk), 24.1 (CH₂^Alk) ppm; **IR (neat):** $\tilde{V}$ = 3064, 3025, 2951, 2875, 1637, 1542, 1440, 1370, 1348, 1161, 937, 851, 729, 700 cm⁻¹; **HRMS (ESI):** m/z calculated for [M+Na]⁺ (C₂₁H₂₅N₃O₅SNa) 454.1407, found 454.1412.

2.1k *N*-Methyl-*N*-(1-oxo-4-phenyl-1-(pyrrolidin-1-yl)butan-2-yl)naphthalene-2-sulfonamide

The product was prepared according to general procedure C. Purification by column chromatography on silica gel (EtOAc:heptane = 1:2) yielded the product (60.8 mg, 70%) as a pale yellow solid.

^{1}H NMR (400 MHz, CDCl$_3$): δ = 8.29 (d, J = 1.6 Hz, 1H^{28}), 7.97–7.89 (m, 3H^{Ar}), 7.69–7.59 (m, 3H^{Ar}), 7.28–7.15 (m, 3H^{Ar}), 7.05–7.00 (m, 2H^{Ar}), 4.66 (dd, J = 9.0, 6.1 Hz, 1H^1), 3.76–3.67 (m, 1H^6), 3.42–3.33 (m, 1H^9), 3.31–3.20 (m, 2H^{6,9}), 3.01 (s, 3H^{11}), 2.63–2.42 (m, 2H^{13}), 2.17–2.05 (m, 1H^{Alk}), 1.94–1.70 (m, 4H^{Alk}), 1.54–1.44 (m, 1H^{Alk}) ppm; **^{13}C NMR (175 MHz, CDCl$_3$):** δ = 167.6 (C^2), 140.7 (C^{Ar}), 136.1 (C^{Ar}), 135.0 (C^{Ar}), 132.3 (C^{Ar}), 129.5 (CH^{Ar}), 129.2 (CH^{Ar}), 128.9 (CH^{Ar}), 128.6 (2CH^{Ar}), 128.6 (2CH^{Ar}), 128.1 (CH^{Ar}), 127.7 (CH^{Ar}), 126.4 (CH^{Ar}), 122.6 (CH^{Ar}), 56.4 (CH^1), 46.6 (CH$_2^6$), 46.1 (CH$_2^9$), 32.3 (CH$_2^{Alk}$), 30.7 (CH$_3^{11}$), 29.7 (CH$_2^{Alk}$), 26.3 (CH$_2^{Alk}$), 24.2 (CH$_2^{Alk}$) ppm; **IR (neat):** $\tilde{V}$ = 3058, 3025, 2969, 2876, 1642, 1447, 1336, 1159, 933, 751, 702, 651 cm^{-1}; **HRMS (ESI):** m/z calculated for [M+H]$^+$ (C$_{25}$H$_{29}$N$_2$O$_3$S) 437.1893, found 437.1892.

2.11 *N*-Methyl-*N*-(1-oxo-4-phenyl-1-(pyrrolidin-1-yl)butan-2-yl)pyridine-2-sulfonamide

The product was prepared according to general procedure C. Purification by column chromatography on silica gel (EtOAc:heptane = 1:2) yielded the product (57.0 mg, 74%) as a pale yellow liquid.

^{1}H NMR (400 MHz, CDCl$_3$): δ = 8.66–8.63 (m, 1H^{27}), 7.92–7.83 (m, 2H^{25,26}), 7.49–7.43 (m, 1H^{Ar}), 7.31–7.24 (m, 2H^{Ar}), 7.21–7.12 (m, 3H^{Ar}), 4.81 (dd, J = 8.0, 6.8 Hz, 2H^1), 3.88–3.79 (m, 1H^6), 3.45–3.32 (m, 2H^9), 3.30–3.22 (m, 1H^6), 3.04 (s, 3H^{11}), 2.67–2.52 (m, 2H^{Alk}), 2.20–2.09 (m, 1H^{Alk}), 1.98–1.77 (m, 5H^{Alk}), 1.60 (bs, 1H^{Alk}) ppm; **^{13}C NMR (100 MHz, CDCl$_3$):** δ = 168.2 (C^2), 157.2 (C^{Ar}), 150.1 (CH^{Ar}), 141.0 (C^{Ar}), 137.9 (CH^{Ar}), 128.6 (CH^{Ar}), 128.6 (CH^{Ar}), 126.7 (CH^{Ar}), 126.3 (CH^{Ar}), 122.8 (CH^{Ar}), 56.9 (CH^1), 46.6 (CH$_2^6$), 46.1 (CH$_2^9$), 32.2 (CH$_2^{Alk}$), 30.9 (CH$_3^{11}$), 30.8 (CH$_2^{Alk}$), 26.3 (CH$_2^{Alk}$), 24.4 (CH$_2^{Alk}$). **IR (neat):** $\tilde{V}$ = 3025, 2951, 2874, 1640, 1577, 1449, 1428, 1340, 1172, 1083, 937, 778, 744, 700, 589 cm^{-1}; **HRMS (ESI):** m/z calculated for [M+Na]$^+$ (C$_{20}$H$_{25}$N$_3$O$_3$SNa) 410.1509, found 410.1507.

2.1n Methyl 8-((*N*,4-dimethylphenyl)sulfonamido)-9-oxo-9-(pyrrolidin-1-yl)nonanoate

The product was prepared according to general procedure C. Purification by column chromatography on silica gel (EtOAc:heptane = 1:1) yielded the product (55.8 mg, 64%) as a pale yellow liquid.

^{1}H NMR (400 MHz, CDCl$_3$): δ = 7.63 (d, J = 8.3 Hz, 2H^{24,28}), 7.28 (d, J = 8.3 Hz, 2H^{25,27}), 4.64–4.58 (m, 1H^1), 3.86–3.76 (m, 1H^6), 3.66 (s, 3H^{17}), 3.50–3.24 (m, 3H^{6,9}), 2.91 (s, 3H^{11}), 2.42 (s, 3H^{29}), 2.28 (t, J = 7.6 Hz, 2H^{13}), 2.01–1.91 (m, 2H^{Alk}), 1.88–1.69 (m, 3H^{Alk}), 1.63–1.52 (m, 2H^{Alk}), 1.33–1.11 (m, 7H^{Alk}) ppm; **^{13}C NMR (175 MHz, CDCl$_3$)**: δ = 174.2 (C^{14}), 168.3 (C^2), 143.5 (C^{21}), 136.4 (C^{26}), 129.6 (2CH^{24,28}), 127.2 (2CH^{25,27}), 56.9 (CH^1), 51.5 (CH$_3$^{17}), 46.6 (CH$_2$^{6}), 45.9 (CH$_2$^{9}), 34.1 (CH$_2$^{Alk}), 30.3 (CH$_3$^{11}), 28.9 (CH$_2$^{Alk}), 28.2 (CH$_2$^{Alk}), 26.3 (CH$_2$^{Alk}), 26.0 (CH$_2$^{Alk}), 24.9 (CH$_2$^{Alk}), 24.2 (CH$_2$^{Alk}), 21.6 (CH$_3$^{27}) ppm; **IR (neat)**: $\overline{V}$ = 2931, 2860, 1734, 1641, 1437, 1335, 1154, 1088, 815, 713, 652 cm^{-1}; **HRMS (ESI)**: m/z calculated for [M+Na]$^+$ (C$_{22}$H$_{34}$N$_2$O$_5$SNa) 461.2081, found 461.2079.

2.1o *N*-(1,10-Dioxo-1-(pyrrolidin-1-yl)undecan-2-yl)-*N*,4-dimethylbenzene-sul-fonamide

The product was prepared according to general procedure C. Purification by column chromatography on silica gel (EtOAc:heptane = 1:1) yielded the product (60.0 mg, 69%) as a pale yellow liquid.

^{1}H NMR (400 MHz, CDCl$_3$): δ = 7.63 (d, *J* = 8.3 Hz, 2H^{24,28}), 7.28 (d, *J* = 8.3 Hz, 2H^{25,27}), 4.62 (dd, *J* = 8.2, 6.2 Hz, 1H^1), 3.85–3.77 (m, 1H^6), 3.51–3.24 (m, 3H^{6,9}), 2.91 (s, 3H^{11}), 2.42 (s, 3H^{29}), 2.40 (t, *J* = 7.4 Hz, 2H^{13}), 2.13 (s, 3H^{23}), 2.00–1.92 (m, 2H^{Alk}), 1.88–1.79 (m, 2H^{Alk}), 1.79–1.69 (m, 1H^{Alk}), 1.60–1.48 (m, 2H^{Alk}), 1.32–1.12 (m, 9H^{Alk}) ppm; **^{13}C NMR (175 MHz, CDCl$_3$)**: δ = 209.3 (C^{35}), 168.4 (C^2), 143.5 (C^{21}), 136.4 (C^{26}), 129.7 (2CH^{24,28}), 127.3 (2CH^{25,27}), 56.9 (CH^1), 46.7(CH$_2^6$), 46.0 (CH$_2^9$), 43.8 (CH$_2^{14}$), 30.4 (CH$_3^{11}$), 30.0 (CH$_3^{23}$), 29.3 (CH$_2^{Alk}$), 29.2 (CH$_2^{Alk}$), 29.1 (CH$_2^{Alk}$), 28.3 (CH$_2^{Alk}$), 26.3 (CH$_2^{Alk}$), 26.2 (CH$_2^{Alk}$), 24.3 (CH$_2^{Alk}$), 23.9 (CH$_2^{Alk}$), 21.6 (CH$_3^{29}$) ppm; **IR (neat)**: $\overline{V}$ = 2928, 2855, 1711, 1640, 1598, 1441, 1334, 1154, 1088, 911, 814, 730, 651 cm^{-1}; **HRMS (ESI)**: m/z calculated for [M+Na]$^+$ (C$_{23}$H$_{36}$N$_2$O$_4$SNa) 459.2288, found 459.2289.

2.1p *N*-(7-Cyano-1-oxo-1-(pyrrolidin-1-yl)hexan-2-yl)-*N*,4-dimethylbenzenesulfonamide

The product was prepared according to general procedure C. Purification by column chromatography on silica gel (EtOAc:heptane = 1:9-7:3) yielded the product (32.5 mg, 47%) as a pale yellow liquid.

^{1}H NMR (400 MHz, CDCl$_3$): δ = 7.65 (d, *J* = 8.2 Hz, 2H^{24,28}), 7.31 (d, *J* = 8.2 Hz, 2H^{25,27}), 4.62 (dd, *J* = 8.9, 6.1 Hz, 1H^1), 3.85–3.77 (m, 1H^6), 3.51–3.28 (m, 3H^{6,9}), 2.87 (s, 3H^{11}), 2.43 (s, 3H^{29}), 2.28 (t, *J* = 7.1 Hz, 2H^{12}), 2.01–1.93 (m, 2H^{Alk}), 1.89–1.76 (m, 3H^{Alk}), 1.72–1.49 (m, 3H^{Alk}), 1.39–1.31 (m, 2H^{Alk}) ppm; **^{13}C NMR (100 MHz, CDCl$_3$)**: δ = 167.6 (C^2), 143.8 (C^{21}), 136.3 (C^{26}), 129.9 (2CH^{24,28}), 127.3 (2CH^{25,27}), 119.5 (C^{12}), 56.9 (CH^1), 46.8 (CH$_2{}^6$), 46.2 (CH$_2{}^9$), 30.4 (CH$_3{}^{11}$), 27.3 (CH$_2{}^{Alk}$), 26.3 (CH$_2{}^{Alk}$), 25.4 (CH$_2{}^{Alk}$), 25.2 (CH$_2{}^{Alk}$), 24.2 (CH$_2{}^{Alk}$), 21.7 (CH$_3{}^{29}$), 17.1 (CH$_2{}^{10}$) ppm; **IR (neat)**: $\overline{V}$ = 2950, 2875, 2246, 1641, 1444, 1334, 1156, 1088, 816, 714, 653 cm^{-1}; **HRMS (ESI)**: m/z calculated for [M+Na]$^+$ (C$_{19}$H$_{27}$N$_3$O$_3$SNa) 400.1665, found 400.1672.

2.1q *N*-(6-Chloro-1-oxo-1-(pyrrolidin-1-yl)hexan-2-yl)-*N*,4-dimethylbenzenesulfonamide

The product was prepared according to general procedure C. Purification by column chromatography on silica gel (EtOAc:heptane = 1:1) yielded the product (51.2 mg, 66%) as a pale yellow liquid.

^{1}H NMR (400 MHz, CDCl$_3$): δ = 7.65 (d, J = 8.4 Hz, 2H^{24,28}), 7.30 (d, J = 8.4 Hz, 2H^{25,27}), 4.63 (dd, J = 8.8, 6.1 Hz, 1H^1), 3.87–3.78 (m, 1H^6), 3.52–3.27 (m, 5H^{6,9,10}), 2.89 (s, 3H^{11}), 2.42 (s, 3H^{29}), 2.02–1.92 (m, 2H^{Alk}), 1.88–1.60 (m, 5H^{Alk}), 1.40–1.21 (m, 3H^{Alk}) ppm; **^{13}C NMR (100 MHz, CDCl$_3$):** δ = 167.9 (C^2), 143.7 (C^{21}), 136.3 (C^{26}), 129.8 (2CH^{24,28}), 127.3 (2CH^{25,27}), 57.0 (CH^1), 46.8 (CH$_2^6$), 46.1 (CH$_2^9$), 44.6 (CH$_2^{10}$), 32.2 (CH$_2^{10}$), 30.4 (CH$_3^{11}$), 27.4 (CH$_2^{Alk}$), 26.3 (CH$_2^{Alk}$), 24.3 (CH$_2^{Alk}$), 23.6 (CH$_2^{Alk}$), 21.7 (CH$_3^{29}$) ppm; **IR (neat):** $\overline{V}$ = 2951, 2873, 1640, 1441, 1334, 1154, 1088, 929, 912, 815, 713, 652 cm^{-1}; **HRMS (ESI):** m/z calculated for [M+Na]$^+$ (C$_{18}$H$_{27}$ClN$_2$O$_3$SNa) 409.1323, found 409.1330.

2.1r N,4-Dimethyl-N-(1-oxo-1-(pyrrolidin-1-yl)undec-10-en-2-yl)benzene-sulfonamide

The product was prepared according to general procedure C. Purification by column chromatography on silica gel (EtOAc:heptane = 1:19-2:3) yielded the product (39.6 mg, 47%) as a pale yellow liquid.

^{1}H NMR (400 MHz, CDCl$_3$): δ = 7.64 (d, J = 8.3 Hz, 2H^{24,28}), 7.28 (d, J = 8.3 Hz, 2H^{25,27}), 5.86–5.75 (m, 1H^{15}), 5.03–4.91 (m, 2H^{16}), 4.62 (dd, J = 8.0, 6.0 Hz, 1H^1), 3.86–3.78 (m, 1H^6), 3.66 (s, 3H^{17}), 3.50–3.25 (m, 3H^{6,9}), 2.91 (s, 3H^{11}), 2.42 (s, 3H^{29}), 2.07–1.92 (m, 2H^{Alk}), 1.88–1.69 (m, 3H^{Alk}), 1.40–1.12 (m, 11H^{Alk}) ppm; **^{13}C NMR (175 MHz, CDCl$_3$):** δ = 168.5 (C^2), 143.5 (C^{21}), 139.2 (CH^{15}), 136.5 (C^{26}), 129.7 (2CH^{24,28}), 127.3 (2CH^{25,27}), 114.4 (CH$_2^{16}$), 57.0 (CH^1), 46.7 (CH$_2^6$), 46.0 (CH$_2^9$), 33.9 (CH$_2^{14}$), 30.4 (CH$_3^{11}$), 29.4 (CH$_2^{Alk}$), 29.3 (CH$_2^{Alk}$), 29.1 (CH$_2^{Alk}$), 29.0 (CH$_2^{Alk}$), 28.3 (CH$_2^{Alk}$), 26.3 (CH$_2^{Alk}$), 26.2 (CH$_2^{Alk}$), 24.3 (CH$_2^{Alk}$), 21.7 (CH$_3^{29}$) ppm; **IR (neat):** $\overline{V}$ = 3070, 2925, 2855, 1643, 1441, 1338, 1160, 1088, 910, 815, 713, 652 cm^{-1}; **HRMS (ESI):** m/z calculated for [M+Na]$^+$ (C$_{23}$H$_{36}$N$_2$O$_3$SNa) 443.2339, found 443.2336.

2.1s   N,N-Dimethyl-2-((N-methyl-4-nitrophenyl)sulfonamido)-4-phenylbutan-amide

The product was prepared according to general procedure C. Purification by column chromatography on silica gel (EtOAc:heptane = 1:5 – 1:2) yielded the product (60.4 mg, 74%) as a pale yellow liquid.

^{1}**H NMR (400 MHz, CDCl$_3$)**: δ = 8.31 (d, J = 9.0 Hz, 2H^{25,27}), 7.81 (d, J = 9.0 Hz, 2H^{24,28}), 7.36–7.30 (m, 2H^{16,18}), 7.28–7.22 (m, 1H^{17}), 7.16–7.10 (m, 2H^{15,19}), 4.83 (apt, J = 7.5 Hz, 1H^1) 3.01 (s, 3H^6), 2.99 (s, 3H^6), 2.86 (s, 3H^{11}), 2.67–2.48 (m, 2H^{13}), 2.17–2.07 (m, 1H^3), 1.63–1.53 (m, 1H^3) ppm; ^{13}**C NMR (100 MHz, CDCl$_3$)**: δ = 169.3 (C^2), 150.2 (C^{26}), 144.7 (C^{Ar}), 140.3 (C^{Ar}), 128.8 (2CH^{Ar}), 128.7 (2CH^{Ar}), 128.5 (2CH^{Ar}), 126.7 (CH^{17}), 124.3 (2CH^{Ar}), 54.4 (CH^1), 37.3 (CH$_3^6$), 36.0 (CH$_3^9$), 32.1 (CH$_2^{Alk}$), 30.7 (CH$_3^{Alk}$), 30.5 (CH$_2^{Alk}$) ppm; **IR (neat)**: $\bar{V}$ = 3103, 3061, 3026, 2933, 2866, 1645, 1528, 1346, 1161, 932, 854, 736, 697, 606 cm^{-1}; **HRMS (ESI)**: m/z calculated for [M+Na]$^+$ (C$_{19}$H$_{23}$N$_3$O$_5$SNa) 428.1251, found 428.1248.

2.1t   N-Methyl-4-nitro-N-(1-oxo-3-phenyl-1-(pyrrolidin-1-yl)propan-2-yl)benzenesulfonamide

The product was prepared according to general procedure C. Purification by column chromatography on silica gel (EtOAc:heptane = 1:3) yielded the product (60.0 mg, 72%) as a pale yellow solid.

¹H NMR (400 MHz, CDCl₃): δ = 8.25 (d, J = 9.0 Hz, 2H[25,27]), 7.78 (d, J = 9.0 Hz, 2H[24,28]), 7.27–7.21 (m, 3H[Ar]), 7.14–7.10 (m, 2H[Ar]), 4.94 (dd, J = 9.2, 6.2 Hz, 1H[1]), 3.65–3.57 (m, 1H[6]), 3.33–3.12 (m, 3H[6,9]), 3.13 (s, 3H[11]), 2.87–2.79 (m, 1H[13]), 2.71 (dd, J = 13.3, 6.2 Hz, 1H[13]), 1.90–1.64 (m, 4H[Alk]) ppm; **¹³C NMR (100 MHz, CDCl₃):** δ = 167.7 (C[2]), 150.0 (C[26]), 144.9 (C[Ar]), 136.2 (C[Ar]), 129.3 (2CH[Ar]), 128.9 (2CH[Ar]), 128.4 (2CH[Ar]), 127.3 (CH[23]), 124.3 (CH[Ar]), 58.3 (CH[1]), 46.5 (CH₂[6]), 45.8 (CH₂[9]), 36.0 (CH₂[3]), 31.0 (CH₃[19]), 26.1 (CH₂[7]), 24.2 (CH₂[8]) ppm; **IR (neat):** $\tilde{V}$ = 3104, 3063, 3029, 2876, 1640, 1528, 1447, 1347, 1162, 942, 853, 738, 699, 609 cm⁻¹; **HRMS (ESI):** m/z calculated for [M+Na]⁺ (C₂₀H₂₃N₃O₅SNa) 440.1244, found 440.1251.

2.1u *N*,4-Dimethyl-*N*-(1-oxo-1-(pyrrolidin-1-yl)butan-2-yl)benzene-sulfonamide

The product was prepared according to general procedure C. Purification by column chromatography on silica gel (EtOAc:heptane = 1:2) yielded the product (37.1 mg, 57%) as a pale yellow liquid.

¹H NMR (400 MHz, CDCl₃): δ = 7.64 (d, J = 8.2 Hz, 2H[17,21]), 7.28 (d, J = 8.2 Hz, 2H[18,20]), 4.57 (dd, J = 8.4, 6.5 Hz, 1H[1]), 3.86–3.79 (m, 1H[6]), 3.50–3.37 (m, 2H[6,9]), 3.34–3.26 (m, 1H[9]), 2.92 (s, 3H[11]), 2.42 (s, 3H[22]), 2.00–1.92 (m, 2H[7]), 1.88–1.75 (m, 3H[2x8,3]), 1.39–1.28 (m, 1H[3]), 0.84 (t, J = 7.5 Hz, 3H[13]) ppm; **¹³C NMR (125 MHz, CDCl₃):** δ = 168.4 (C[2]), 143.5 (C[14]), 136.5 (C[19]), 129.7 (2CH[17,21]), 127.3 (2CH[18,20]), 58.5 (CH[1]), 46.7 (CH[6]), 46.0 (CH[9]), 30.3 (CH₃[11]), 26.3 (CH₂[Alk]), 24.3 (CH₂[Alk]), 21.7 (CH₂[Alk]), 21.6 (CH₃[22]), 10.8 (CH₃[13]) ppm; **IR (neat):** $\tilde{V}$ = 2969, 2877, 1642, 1444, 1336, 1163, 1151, 1088, 955, 812, 714, 652 cm⁻¹; **HRMS (ESI):** m/z calculated for [M+Na]⁺ (C₁₆H₂₄N₂O₃SNa) 347.1400, found 347.1401.

2.1v *N*-Methyl-4-nitro-*N*-(1-oxo-1-(pyrrolidin-1-yl)butan-2-yl)benzenesulfon-amide

The product was prepared according to general procedure C. Purification by column chromatography on silica gel (EtOAc:heptane = 2:3) yielded the product (40.3 mg, 55%) as a pale yellow liquid.

^{1}H NMR (400 MHz, CDCl$_3$): δ = 8.33 (d, J = 9.0 Hz, 2H[17,21]), 7.93 (d, J = 9.0 Hz, 2H[18,20]), 4.60 (t, J = 7.5 Hz, 1H[1]), 3.80–3.72 (m, 1H[6]), 3.51–3.33 (m, 2H[6,9]), 3.30–3.21 (m, 1H[9]), 3.02 (s, 3H[11]), 2.04–1.94 (m, 2H[7]), 1.91–1.76 (m, 3H[2x8,3]), 1.54–1.43z (m, 1H[3]), 0.90 (t, J = 7.3 Hz, 3H[13]) ppm; **^{13}C NMR (175 MHz, CDCl$_3$):** δ = 168.3 (C[2]), 150.1 (C[19]), 145.1 (C[14]), 128.4 (2CH[17,21]), 124.3 (2CH[18,20]), 58.7 (CH[1]), 46.7 (CH$_2$[6]), 46.0 (CH$_2$[9]), 30.6 (CH$_3$[11]), 26.4 (CH$_2$[Alk]), 24.3 (CH$_2$[Alk]), 22.3 (CH$_2$[Alk]), 10.9 (CH$_3$[13]) ppm; **IR (neat):** $\tilde{\nu}$ = 3103, 2971, 2876, 1642, 1529, 1446, 1348, 1310, 1155, 1087, 739, 606 cm^{-1}; **HRMS (ESI):** m/z calculated for [M+Na]$^+$ (C$_{15}$H$_{21}$N$_3$O$_5$SNa) 378.1094, found 378.1092.

2.1y Methyl 9-((*R*)-2-benzhydrylpyrrolidin-1-yl)-8-((*N*,4-dimethylphenyl)sulfon-amido)-9-oxononanoate

The product was prepared according to general procedure C. Purification by column chromatography on silica gel (EtOAc:heptane = 1:5–1:2) yielded the product (34.4 mg, 28%) as a white solid. Isolated as a mixture of diasteromers in a ratio of 1:1.3, only the major one is reported for the ^{1}H NMR. Both diastereomers are reported for the ^{13}C NMR (maj = major, min = minor) and the IR.

^{1}**H NMR (700 MHz, CDCl$_3$):** δ = 7.62 (d, *J* = 8.3 Hz, 2H^{24,28}), 7.48–7.12 (m, 12H^{Ar}), 5.51 (dd, *J* = 10.8, 6.1 Hz, 1H^9), 4.06 (t, *J* = 7.1 Hz, 1H^1), 3.91–3.58 (m, 2H^{6,18}), 3.68 (s, 3H^{17}), 3.50 (m, 1H^6), 2.62 (s, 3H^{11}), 2.42 (s, 3H^{29}), 2.25 (t, *J* = 7.4 Hz, 2H^{13}), 2.00–1.85 (m, 4H^{7,8}), 1.51–1.47 (m, 2H^{12}), 1.20–0.78 (m, 5H^{Alk}), 0.65–0.59 (m, 1H^{Alk}), 0.46–0.39 (m, 1H^{Alk}), 0.30–0.23 (m, 1H^{Alk}) ppm. ^{13}**C NMR (175 MHz, CDCl$_3$):** 174.3 (C^{14}), 168.3 (C^{2,maj}), 168.0 (C^{2,min}), 143.7 (C^{21}), 143.6 (C^{21}), 142.2 (C^{Ar}), 142.1 (C^{Ar}), 142.1 (C^{Ar}), 141.6 (C^{Ar}), 136.4 (C^{26}), 129.8 (2C^{Ar}), 129.8 (2C^{Ar}), 129.5 (2C^{Ar}), 129.4 (2C^{Ar}), 129.1 (2C^{Ar}), 129.0 (2C^{Ar}), 129.0 (2C^{Ar}), 128.7 (2C^{Ar}), 128.6 (2C^{Ar}), 128.2 (2C^{Ar}), 127.5 (2C^{Ar}), 127.5 (2C^{Ar}), 127.0 (C^{Ar}), 126.9 (C^{Ar}), 126.8 (C^{Ar}), 126.3 (C^{Ar}), 60.8 (CH^{9,maj}), 60.2 (CH^{min}), 57.6 (CH^{min}), 57.1 (CH^{1,maj}), 54.8 (CH^{18,maj}), 52.9 (CH^{min}), 51.6 (CH$_3$^{17}), 46.9 (CH$_2$^{min}), 44.8 (CH$_2$^{6,maj}), 34.2 (CH$_2$^{13,maj}), 34.1 (CH$_2$^{13,min}), 30.3 (CH$_2$^{Alk}), 30.0 (CH$_3$^{11,maj}), 29.9 (CH$_3$^{11,min}), 29.2 (CH$_2$^{Alk}), 29.1 (CH$_2$^{Alk}), 29.0 (CH$_2$^{Alk}), 27.9 (CH$_2$^{Alk}), 26.9 (CH$_2$^{Alk}), 26.9 (CH$_2$^{Alk}), 26.4 (CH$_2$^{Alk}), 25.6 (CH$_2$^{Alk}), 25.0 (CH$_2$^{Alk}), 24.9 (CH$_2$^{Alk}), 23.6 (CH$_2$^{Alk}), 21.9 (CH$_3$^{29,min}), 21.7 (CH$_3$^{29,maj}), 21.1 (CH$_2$^{Alk}) ppm. **IR (neat):** $\tilde{V}$ = 3056, 3028, 2927, 2856, 1733, 1634, 1599, 1493, 1434, 1336, 1265, 1159, 1087, 814, 732 cm^{-1}; **HRMS (ESI):** m/z calculated for [M+Na]$^+$ (C$_{35}$H$_{44}$N$_2$O$_5$SNa) 627.2863, found 627.2862.

2.1z 2-Amino-4-phenyl-1-(pyrrolidin-1-yl)butan-1-one

To 32 mg (0.077 mmol) **1d** in 0.5 mL MeCN 25.3 mg (0.23 mmol) thiophenol and 42.4 mg (0.307 mmol) freshly ground K_2CO_3 was added. The mixture was heated to 50 °C for 40 h. Then, the solvent was removed under reduced pressure and the crude product was purified by column chromatography on silica gel (EtOAc:(99% MeOH, 1% NH$_4$OH) = 9:1) to yield amine **1z** (16.6 mg, 93%) as a pale yellow liquid.

^{1}H NMR (400 MHz, CDCl$_3$): δ = 7.31–7.25 (m, 2H^{Ar}), 7.23–7.16 (m, 3H^{Ar}), 3.54–3.37 (m, 3H^{1,7}), 3.34–3.26 (m, 1H^{10}), 3.19–3.10 (m, 1H^{10}), 2.86–2.67 (m, 2H^{11}), 1.96–1.62 (m, 8H^{NH,3,8,9}), ppm; **^{13}C NMR (100 MHz, CDCl$_3$)**: δ = 174.1 (C^2), 141.7 (C^{12}), 128.7 (2CH^{Ar}), 128.6 (2CH^{Ar}), 126.1 (CH^{15}), 52.4 (CH^1), 46.1 (CH$_2{}^7$), 46.0 (CH$_2{}^{10}$), 36.9 (CH$_2{}^{Alk}$), 32.2 (CH$_2{}^{Alk}$), 26.2 (CH$_2{}^{Alk}$), 24.2 (CH$_2{}^{Alk}$) ppm; **IR (neat)**: $\tilde{V}$ = 3363, 3060, 3025, 2950, 2873, 1631, 1449, 1370, 1342, 752, 701 cm^{-1}; **HRMS (ESI)**: m/z calculated for [M+H]$^+$ (C$_{14}$H$_{21}$N$_2$O) 233.1648, found 233.1647.

2.2a 2-(4-Nitrophenoxy)-4-phenyl-1-(pyrrolidin-1-yl)butan-1-one

The product was prepared according to general procedure C. Purification by column chromatography on silica gel (EtOAc:heptane = 1:1) yielded the product (56.0 mg, 79%) as a pale yellow liquid.

^{1}H NMR (400 MHz, CDCl$_3$): δ = 8.16 (d, J = 9.3 Hz, 2H^{25,27}), 7.29–7.12 (m, 5H^{Ar}), 6.87 (d, J = 9.3 Hz, 2H^{24,28}), 4.66 (dd, J = 9.1, 4.0 Hz, 1H^1), 3.53–3.37 (m,

$3H^{6,9}$), 3.25–3.14 (m, $1H^6$), 2.96–2.87 (m, $1H^{13}$), 2.85–2.74 (m, $1H^{13}$), 2.38–2.26 (m, $1H^3$), 2.23–2.13 (m, $1H^3$), 1.94–1.68 (m, $4H^{7,8}$), ppm; **^{13}C NMR (100 MHz, CDCl$_3$)**: δ = 168.1 (C^2), 162.8 (C^{21}), 142.2 (C^{26}), 140.5 (C^{14}), 128.7 (2CH^{Ar}), 128.7 (2CH^{Ar}), 126.5 (CH^{17}), 126.2 (2CH^{Ar}), 115.0 (2CH^{Ar}), 77.8 (CH^1), 46.8 (CH$_2$^{6}), 46.0 (CH$_2$^{7}), 33.3 (CH$_2$^{Alk}), 31.7 (CH$_2$^{Alk}), 26.6 (CH$_2$^{Alk}), 23.5 (CH$_2$^{Alk}) ppm; **IR (neat)**: $\overline{\nu}$ = 3082, 3027, 2956, 2878, 1652, 1590, 1512, 1494, 1441, 1338, 1256, 1110, 845, 751, 655 cm^{-1}; **HRMS (ESI)**: m/z calculated for [M+Na]$^+$ (C$_{20}$H$_{22}$N$_2$O$_4$Na) 377.1472, found 377.1475.

2.2n 2-(Benzyloxy)-*N,N*-dimethyl-4-phenylbutanamide

The product was prepared according to general procedure C. Purification by column chromatography on silica gel (EtOAc:heptane = 1:3) yielded the product (44.0 mg, 74%) as a pale yellow liquid.

    ^{1}H NMR (400 MHz, CDCl$_3$): δ = 7.40–7.11 (m, 10H^{Ar}), 6.87 (d, *J* = 9.3 Hz, 2H^{24,28}), 4.63 (d, *J* = 11.6 Hz, 1H^{21}), 4.35 (d, *J* = 11.6 Hz, 1H^{21}), 4.16 (dd, *J* = 9.1, 4.2 Hz, 1H^1), 2.94 (s, 3H^6), 2.92 (s, 3H^9), 2.89–2.81 (m, 1H^{13}), 2.74–2.65 (m, 1H^{13}), 2.20–2.09 (m, 1H^3), 2.02–1.92 (m, 1H^3) ppm; **^{13}C NMR (100 MHz, CDCl$_3$)**: δ = 171.7 (C^2), 141.5 (C^{Ar}), 137.9 (C^{Ar}), 128.7 (2CH^{Ar}), 128.6 (2CH^{Ar}), 128.5 (2CH^{Ar}), 128.3 (2CH^{Ar}), 128.0 (2CH^{Ar}), 126.1 (2CH^{Ar}), 77.5 (CH^1), 71.5 (CH$_2$^{21}), 36.52 (CH$_3$^{6}), 36.2 (CH$_3$^{9}), 33.9 (CH$_2$^{Alk}), 31.9 (CH$_2$^{Alk}) ppm; **IR (neat)**: $\overline{\nu}$ = 3060, 3026, 2925, 2861, 1639, 1494, 1452, 1397, 1343, 1257, 1097, 981, 734, 697 cm^{-1}; **HRMS (ESI)**: m/z calculated for [M+Na]$^+$ (C$_{19}$H$_{23}$NO$_2$Na) 320.1621, found 320.1621.

2.2o 1-(Dimethylamino)-1-oxopentan-2-yl formate

The product was prepared according to general procedure C with the difference that no sulfonamide or alcohol was added. Purification by column chromatography on silica gel (EtOAc:heptane = 1:1) yielded the product (21.1 mg, 26%) as a pale yellow liquid.

^{1}H NMR (400 MHz, CDCl$_3$): δ = 8.08 (s, 1H^8), 5.40 (dd, J = 4.3, 8.6 Hz, 1H^1), 3.08 (s, 3H^{10}), 3.96 (s, 3H^{11}), 1.91–1.80 (m, 1H^3), 1.77–1.66 (m, 1H^3), 1.56–1.35 (m, 2H^7), 0.95 (t, J = 7.3 Hz, 3H^{12}) ppm; **^{13}C NMR (175 MHz, CDCl$_3$):** δ = 169.4 (C^2), 160.7 (CH^8), 69.9 (CH^1), 37.0 (CH$_3$^{10}), 36.1 (CH$_3$^{10}), 33.1 (CH$_2$^{3}), 18.6 (CH$_2$^{7}), 13.8 (CH$_3$^{10}) ppm; **IR (neat):** $\tilde{V}$ = 2960, 2934, 2874, 1720, 1652, 1500, 1462, 1402, 1259, 1163, 1108, 1059, 852, 631 cm^{-1}; **HRMS (ESI):** m/z calculated for [M+Na]$^+$ (C$_8$H$_{15}$NO$_3$Na) 196.0944, found 196.0942.

2.2p 2-Hydroxy-N,N-dimethylpentanamide

The product was prepared according to general procedure C with the difference that no sulfonamide or alcohol was added. Purification by column chromategraphy (EtOAc:heptane = 1:1) yielded the product (36.3 mg, 42%) as a pale yellow liquid. All spectroscopic properties are in good accordance with reported data.[51]

General procedure D for the synthesis of α-arylated amides 2.3a–c

To a solution of the amide (1 eq.) in MeCN (0.1 M) was added tetrapropyl ammonium hydroxide (40% (aq.), 2 eq.) The solution was stirred for 16 h. Then it was washed with saturated NH$_4$Cl solution and brine. The organic layer was dried with MgSO$_4$, the solvent evaporated and the crude product purified by column chromatography on silica gel.

2.3a 2-(Methylamino)-2-(4-nitrophenyl)-4-phenyl-1-(pyrrolidin-1-yl)butan-1-one

The product was prepared according to general procedure D from amide **2.1b** (63 mg, 0.15 mmol). Purification by column chromatography on silica gel (EtOAc:heptane = 1:1) yielded the product (24.0 mg, 45%) as a pale yellow liquid.

^{1}H NMR (400 MHz, CDCl$_3$): δ = 8.21 (d, J = 9.0 Hz, 2H^{9,11}), 7.57 (d, J = 9.0 Hz, 2H^{8,12}), 7.25–7.20 (m, 2H^{Ar}), 7.19–7.13 (m, 1H^{23}), 7.12–7.06 (m, 2H^{Ar}), 3.53 (bs, 2H^{Alk}), 3.19 (bs, 1H^{Alk}), 2.70 (bs, 1H^{Alk}), 2.62–2.53 (m, 1H^{Alk}), 2.42–2.33 (m, 2H^{Alk}), 2.28 (s, 3H^{13}), 1.78–1.49 (m, 4H^{Alk}) ppm; **^{13}C NMR (150 MHz, CDCl$_3$)**: δ = 170.6 (C^{18}), 150.0 (C^{Ar}), 147.0 (C^{Ar}), 141.7 (C^{Ar}), 128.5 (2CH^{Ar}), 128.4 (2CH^{Ar}), 127.3 (2CH^{Ar}), 126.1 (CH^{23}), 123.8 (2CH^{Ar}), 68.5 (C^{19}), 47.8 (CH$_2$^{1}), 46.6 (CH$_2$^{4}), 35.4 (CH$_2$^{Alk}), 29.4 (CH$_2$^{Alk}), 29.2 (CH$_2$^{13}), 26.8 (CH$_3$^{Alk}), 23.2 (CH$_2$^{Alk}) ppm; **IR (neat)**: $\widetilde{\nu}$ = 2946, 2876, 2798, 1625, 1603, 1519, 1401, 1346, 1109, 854 cm^{-1}; **HRMS (ESI)**: m/z calculated for [M+H]$^+$ (C$_{21}$H$_{26}$N$_3$O$_3$) 368.1969, found 368.1971.

2.3b 2-(Methylamino)-2-(4-nitrophenyl)-3-phenyl-1-(pyrrolidin-1-yl)propan-1-one

The product was prepared according to general procedure D from amide **2.1t** (417.5 mg, 1 mmol). Purification by column chromatography on silica gel (EtOAc:heptane = 3:2) yielded the product (159.0 mg, 45%) as a pale yellow solid.

^{1}H NMR (600 MHz, CDCl$_3$): δ = 8.10 (d, J = 9.0 Hz, 2H^{9,11}), 7.31–7.24 (m, 2H^{Ar}), 7.14–7.06 (m, 3H^{Ar}), 6.52 (d, J = 9.0 Hz, 2H^{8,12}), 3.71 (d, J = 14.3 Hz, 1H^{20}), 3.67–3.62 (m, 1H^1), 3.60–3.49 (m, 2H^{1,4}), 3.24 (d, J = 14.3 Hz, 1H^{20}), 2.49 (s, 3H^{13}), 2.34–2.36 (m, 1H^4), 1.79–1.52 (m, 4H^{2,3}), ppm; **^{13}C NMR (150 MHz, CDCl$_3$):** δ = 169.8 (C^{18}), 149.8 (C^{Ar}), 146.7 (C^{26}), 135.9 (C^{Ar}), 130.4 (CH^{Ar}), 128.1 (CH^{Ar}), 127.2 (CH^{Ar}), 126.8 (CH^{Ar}), 123.3 (CH^{Ar}), 69.53 (C^{19}), 47.6 (CH$_2$^{1}), 46.4 (CH$_2$^{4}), 41.1 (CH$_2$^{Alk}), 30.1 (CH$_3$^{13}), 26.7 (CH$_2$^{Alk}), 23.3 (CH$_2$^{Alk}) ppm; **IR (neat):** $\overline{V}$ = 3028, 2947, 2875, 2802, 1629, 1599, 1518, 1451, 1398, 1345, 1107, 850, 705 cm^{-1}; **HRMS (ESI):** m/z calculated for [M+H]$^+$ (C$_{20}$H$_{24}$N$_3$O$_3$) 354.1812, found 354.1816.

2.3c *N,N*-Dimethyl-2-(methylamino)-2-(4-nitrophenyl)-4-phenylbutanamide

The product was prepared according to general procedure D from amide **2.1s** (39 mg, 0.096 mmol). Purification by column chromatography on silica gel (EtOAc:heptane = 1:2) yielded the product (17.0 mg, 51%) as a colourless liquid.

^{1}H NMR (400 MHz, CDCl$_3$): δ = 8.23 (d, J = 9.0 Hz, 2H^{9,11}), 7.53 (d, J = 9.0 Hz, 2H^{8,12}), 7.26–7.20 (m, 2H^{Ar}), 7.19–7.13 (m, 1H^{23}), 7.10–7.04 (m, 2H^{Ar}), 2.91 (bs, 3H^1), 2.73 (bs, 3H^4), 2.65–2.53 (m, 1H^{Alk}), 2.39–2.26 (m, 2H^{21}), 2.30 (s, 3H^{13}), 2.25–2.12 (m, 1H^{20}), 1.97 (bs, 1H^{NH}) ppm; **^{13}C NMR (100 MHz, CDCl$_3$):** δ = 171.7 (C^{18}), 150.4 (C^{Ar}), 146.9 (C^{Ar}), 141.6 (C^{Ar}), 128.6 (2CH^{Ar}), 128.4 (2CH^{Ar}), 126.8 (2CH^{Ar}), 126.1 (CH^{23}), 123.9 (2CH^{Ar}), 68.5 (C^{19}), 37.6 (CH$_2$^{Alk}), 36.5 (CH$_2$^{1,4}), 29.5 (CH$_2$^{Alk}), 29.4 (CH$_2$^{13}) ppm; **IR (neat):** $\overline{V}$ = 2928, 2857, 1666, 1634, 1601, 1518, 1385, 1345, 1256, 1091, 855, 701, 659 cm^{-1}; **HRMS (ESI):** m/z calculated for [M+H]$^+$ (C$_{19}$H$_{24}$N$_3$O$_3$) 342.1813, found 342.1813.

References

(1) Claisen, L.; Claparède, A. Condensationen von Ketonen Mit Aldehyden. *Ber. Dtsch. Chem. Ges.* **1881**, *14*, 2460–2468.

(2) Claisen, L. Ueber Die Einführung von Säureradicalen in Ketone. *Ber. Dtsch. Chem. Ges.* **1887**, *20*, 655–657.

(3) Seebach, D. Methods of Reactivity Umpolung. *Angew. Chem. Int. Ed. Engl.* **1979**, *18*, 239–258.

(4) Seebach, D.; Corey, E. J. Generation and Synthetic Applications of 2-Lithio-1,3-Dithianes. *J. Org. Chem.* **1975**, *40* (2), 231–237.

(5) Ganguly, N.; Barik, S. A Facile Mild Deprotection Protocol for 1,3-Dithianes and 1,3-Dithiolanes with 30% Hydrogen Peroxide and Iodine Catalyst in Aqueous Micellar System. *Synthesis* **2009**, 8, 1393–1399.

(6) Adlington, R. M.; Baldwin, J. E.; Bottaro, J. C.; Perry, M. W. D. Azo Anions in Synthesis. t-Butylhydrazones as Acyl-Anion Equivalents. *J. Chem. Soc. Chem. Commun.* **1983**, 18, 1040–1041.

(7) Wöhler; Liebig. Untersuchungen Über Das Radikal Der Benzoesäure. *Ann. Phar.* **1832**, *3*, 249–282.

(8) Stetter, H. Catalyzed Addition of Aldehydes to Activated Double Bonds? A New Synthetic Approach. *Angew. Chem. Int. Ed. Engl.* **1976**, *15*, 639–647.

(9) Breslow, R. On the Mechanism of Thiamine Action. IV. [1] Evidence from Studies on Model Systems. *J. Am. Chem. Soc.* **1958**, *80*, 3719–3726.

(10) Enders, D.; Breuer, K.; Runsink, J.; Teles, J. H. The First Asymmetric Intramolecular Stetter Reaction. Preliminary Communication. *Helv. Chim. Acta* **1996**, *79*, 1899–1902.

(11) Enders, D.; Han, J.; Henseler, A. Asymmetric Intermolecular Stetter Reactions Catalyzed by a Novel Triazolium Derived N-Heterocyclic Carbene. *Chem. Commun.* **2008**, 34, 3989–3991.

(12) Antranikian, G. *Angewandte Mikrobiologie*; Springer: Berlin, 2006.

(13) Beeson, T. D.; Mastracchio, A.; Hong, J.-B.; Ashton, K.; MacMillan, D. W. C. Enantioselective Organocatalysis Using SOMO Activation. *Science* **2007**, *316*, 582–585.

(14) Narasaka, K.; Okauchi, T.; Tanaka, K.; Murakami, M. Generation of Cation Radicals from Enamines and Their Reactions with Olefins. *Chem. Lett.* **1992**, *21*, 2099–2102.

(15) Mizar, P.; Wirth, T. Flexible Stereoselective Functionalizations of Ketones through Umpolung with Hypervalent Iodine Reagents. *Angew. Chem. Int. Ed.* **2014**, *53*, 5993–5997.

(16) Shneider, O. S.; Pisarevsky, E.; Fristrup, P.; Szpilman, A. M. Oxidative Umpolung α-Alkylation of Ketones. *Org. Lett.* **2015**, *17*, 282–285.

(17) Miyoshi, T.; Miyakawa, T.; Ueda, M.; Miyata, O. Nucleophilic α-Arylation and α-Alkylation of Ketones by Polarity Inversion of N-Alkoxyenamines: Entry to the Umpolung Reaction at the α-Carbon Position of Carbonyl Compounds. *Angew. Chem. Int. Ed.* **2011**, *50*, 928–931.

(18) Quiclet-Sire, B.; Tölle, N.; Zafar, S. N.; Zard, S. Z. Oxime Derivatives as α-Electrophiles. From α-Tetralone Oximes to Tetracyclic Frameworks. *Org. Lett.* **2011**, *13*, 3266–3269.

(19) Ruider, S. A.; Maulide, N. Strong Bonds Made Weak: Towards the General Utility of Amides as Synthetic Modules. *Angew. Chem. Int. Ed.* **2015**, *54*, 13856–13858.

(20) Falmagne, J.-B.; Escudero, J.; Taleb-Sahraoui, S.; Ghosez, L. Cyclobutanone and Cyclobutenone Derivatives by Reaction of Tertiary Amides with Alkenes or Alkynes. *Angew. Chem. Int. Ed. Engl.* **1981**, *20*, 879–880.

(21) Charette, A. B.; Grenon, M. Spectroscopic Studies of the Electrophilic Activation of Amides with Triflic Anhydride and Pyridine. *Can. J. Chem.* **2001**, *79*, 1694–1703.

(22) Kaiser, D.; Maulide, N. Making the Least Reactive Electrophile the First in Class: Domino Electrophilic Activation of Amides. *J. Org. Chem.* **2016**, *81*, 4421–4428.

(23) Peng, B.; Geerdink, D.; Farès, C.; Maulide, N. Chemoselective Intermolecular α-Arylation of Amides. *Angew. Chem. Int. Ed.* **2014**, *53*, 5462–5466.

(24) Shaaban, S.; Tona, V.; Peng, B.; Maulide, N. Hydroxamic Acids as Chemoselective (*Ortho* -Amino) Arylation Reagents *via* Sigmatropic Rearrangement. *Angew. Chem. Int. Ed.* **2017**, *56*, 10938–10941.

(25) Tona, V.; De La Torre, A.; Padmanaban, M.; Ruider, S.; González, L.; Maulide, N. Chemo- and Stereoselective Transition-Metal-Free Amination of Amides with Azides. *J. Am. Chem. Soc.* **2016**, *138*, 8348–8351.

(26) Da Costa, R.; Gillard, M.; Falmagne, J. B.; Ghosez, L. Aβ-Dehydrogenation of Carboxamides. *J. Am. Chem. Soc.* **1979**, *101*, 4381–4383.

(27) Kaiser, D.; de la Torre, A.; Shaaban, S.; Maulide, N. Metal-Free Formal Oxidative C−C Coupling by In Situ Generation of an Enolonium Species. *Angew. Chem. Int. Ed.* **2017**, *56*, 5921–5925.

(28) Di Mauro, G.; Maryasin, B.; Kaiser, D.; Shaaban, S.; González, L.; Maulide, N. Mechanistic Pathways in Amide Activation: Flexible Synthesis of Oxazoles and Imidazoles. *Org. Lett.* **2017**, *19*, 3815–3818.

(29) De La Torre, A.; Kaiser, D.; Maulide, N. Flexible and Chemoselective Oxidation of Amides to α-Keto Amides and α-Hydroxy Amides. *J. Am. Chem. Soc.* **2017**, *139*, 6578–6581.

(30) Kaiser, D.; Teskey, C. J.; Adler, P.; Maulide, N. Chemoselective Intermolecular Cross-Enolate-Type Coupling of Amides. *J. Am. Chem. Soc.* **2017**, *139*, 16040–16043.

(31) Holden, C. M.; Greaney, M. F. Modern Aspects of the Smiles Rearrangement. *Chem. Eur. J.* **2017**, *23*, 8992–9008.

(32) Meisenheimer, J. Ueber Reactionen Aromatischer Nitrokörper. *Justus Liebig's Ann. der Chemie* **1902**, *323*, 205–246.

(33) Wilson, M. W.; Ault-Justus, S. E.; Hodges, J. C.; Rubin, J. R. A Facile Rearrangement of N-Alkyl, N-(0 or p-Nitrophenylsulfonamide)- α-Amino Esters. *Tetrahedron* **1999**, *55*, 1647–1656.

(34) Lupi, V.; Penso, M.; Foschi, F.; Gassa, F.; Mihali, V.; Tagliabue, A. Highly Stereo-selective Intramolecular α-Arylation of Self-Stabilized Non-Racemic Enolates: Synthesis of α-Quaternary α-Amino Acid Derivatives. *Chem. Commun.* **2009**, *33*, 5012.

(35) Smyslová, P.; Kisseljova, K.; Krchňák, V. Base-Mediated Intramolecular C- and N-Arylation of N,N-Disubstituted 2-Nitrobenzenesulfonamides: Advanced Intermedia-tes for the Synthesis of Diverse Nitrogenous Heterocycles. *ACS Comb. Sci.* **2014**, *16*, 500–505.

(36) Lemmerer, M.; Teskey, C. J.; Kaiser, D.; Maulide, N. Regioselective Synthesis of Pyridines by Redox Alkylation of Pyridine N-Oxides with Malonates. *Monatsh. Chem.* **2018**, *149*, 715–719.

(37) Markó, I.; Ronsmans, B.; Hesbain-Frisque, A. M.; Dumas, S.; Ghosez, L.; Ernst, B.; Greuter, H. Intramolecular [2 + 2] Cycloadditions of Ketenes and Keteniminium Salts to Olefins. *J. Am. Chem. Soc.* **1985**, *107*, 2192–2194.

(38) Fukuyama, T.; Jow, C. K.; Cheung, M. 2- and 4-Nitrobenzenesulfonamides: Exceptionally Versatile Means for Preparation of Secondary Amines and Protection of Amines. *Tetrahedron Lett.* 1995, 6373–6374.

(39) Patil, D. V.; Kim, S. W.; Nguyen, Q. H.; Kim, H.; Wang, S.; Hoang, T.; Shin, S. Brønsted Acid Catalyzed Oxygenative Bimolecular Friedel-Crafts-Type Coupling of Ynamides. *Angew. Chem. Int. Ed.* **2017**, *56*, 3670–3674.

(40) Phipps, R. J.; Hamilton, G. L.; Toste, F. D. The Progression of Chiral Anions from Concepts to Applications in Asymmetric Catalysis. *Nat. Chem.* **2012**, *4*, 603–614.

(41) Søren Bertelsen; Mauro Marigo; Sebastian Brandes; Peter Dinér; Jørgensen, K. A. Dienamine Catalysis: Organocatalytic Asymmetric γ-Amination of α,β-Unsaturated Aldehydes. *J. Am. Chem. Soc.* **2006**, 128, 12973–12980.

(42) Zhang, F.; Das, S.; Walkinshaw, A. J.; Casitas, A.; Taylor, M.; Suero, M. G.; Gaunt, M. J. Cu-Catalyzed Cascades to Carbocycles: Union of Diaryliodonium Salts with Alkenes or Alkynes Exploiting Remote Carbocations. *J. Am. Chem. Soc.* **2014**, *136*, 8851–8854.

(43) Kim, I.; Lee, C. Rhodium-Catalyzed Oxygenative Addition to Terminal Alkynes for the Synthesis of Esters, Amides, and Carboxylic Acids. *Angew. Chem. Int. Ed.* **2013**, *52*, 10023–10026.

(44) Feldman, K. S.; Folda, T. S. Studies on the Synthesis of the Alkaloid (-)-Gilbertine *via* Indolidene Chemistry. *J. Org. Chem.* **2016**, *81*, 4566–4575.

(45) Saidykhan, A.; Bowen, R. D.; Gallagher, R. T.; Martin, W. H. C. Intramolecular NC Rearrangements Involving Sulfonamide Protecting Groups. *Tetrahedron Lett.* **2015**, *56*, 66–68.

(46) Gioiello, A.; Rosatelli, E.; Teofrasti, M.; Filipponi, P.; Pellicciari, R. Building a Sulfonamide Library by Eco-Friendly Flow Synthesis. *ACS Comb. Sci.* **2013**, *15*, 235–239.

(47) Chow, S. Y.; Stevens, M. Y.; Odell, L. R. Sulfonyl Azides as Precursors in Ligand-Free Palladium-Catalyzed Synthesis of Sulfonyl Carbamates and Sulfonyl Ureas and Synthesis of Sulfonamides. *J. Org. Chem.* **2016**, *81*, 2681–2691.

(48) Miura, T.; Yamauchi, M.; Kosaka, A.; Murakami, M. Nickel-Catalyzed Regio- and Enantioselective Annulation Reactions of 1,2,3,4-Benzothiatriazine-1,1(2H)-Dioxides with Allenes. *Angew. Chem. Int. Ed.* **2010**, *49*, 4955–4957.

(49) Liang, R.; Li, S.; Wang, R.; Lu, L.; Li, F. N-Methylation of Amines with Methanol Catalyzed by a Cp*Ir Complex Bearing a Functional 2,2'-Bibenzimidazole Ligand. *Org. Lett.* **2017**, 19, 5790–5793.

(50) Wright, S. W.; Hallstrom, K. N. A Convenient Preparation of Heteroaryl Sulfonamides and Sulfonyl Fluorides from Heteroaryl Thiols. *J. Org. Chem.* **2006**, *71*, 1080–1084.

(51) Yao, Y.; Tong, W.; Chen, J. A-Hydroxy Amides from Carbamoylsilane and Aldehydes. *Mendeleev Commun.* **2014**, *24*, 176–177.

Printed in the United States
By Bookmasters